FREE Test Taking Tips DVD Offer

To help us better serve you, we have developed a Test Taking Tips DVD that we would like to give you for FREE. **This DVD covers world-class test taking tips that you can use to be even more successful when you are taking your test.**

All that we ask is that you email us your feedback about your study guide. Please let us know what you thought about it – whether that is good, bad or indifferent.

To get your **FREE Test Taking Tips DVD**, email freedvd@studyguideteam.com with "FREE DVD" in the subject line and the following information in the body of the email:

    a. The title of your study guide.

    b. Your product rating on a scale of 1-5, with 5 being the highest rating.

    c. Your feedback about the study guide. What did you think of it?

    d. Your full name and shipping address to send your free DVD.

If you have any questions or concerns, please don't hesitate to contact us at freedvd@studyguideteam.com.

Thanks again!

AFOQT Study Guide

AFOQT Test Prep Book Team

Table of Contents

Quick Overview

As you draw closer to taking your exam, effective preparation becomes more and more important. Thankfully, you have this study guide to help you get ready. Use this guide to help keep your studying on track and refer to it often.

This study guide contains several key sections that will help you be successful on your exam. The guide contains tips for what you should do the night before and the day of the test. Also included are test-taking tips. Knowing the right information is not always enough. Many well-prepared test takers struggle with exams. These tips will help equip you to accurately read, assess, and answer test questions.

A large part of the guide is devoted to showing you what content to expect on the exam and to helping you better understand that content. Near the end of this guide is a practice test so that you can see how well you have grasped the content. Then, answers explanations are provided so that you can understand why you missed certain questions.

Don't try to cram the night before you take your exam. This is not a wise strategy for a few reasons. First, your retention of the information will be low. Your time would be better used by reviewing information you already know rather than trying to learn a lot of new information. Second, you will likely become stressed as you try to gain large amount of knowledge in a short amount of time. Third, you will be depriving yourself of sleep. So be sure to go to bed at a reasonable time the night before. Being well-rested helps you focus and remain calm.

Be sure to eat a substantial breakfast the morning of the exam. If you are taking the exam in the afternoon, be sure to have a good lunch as well. Being hungry is distracting and can make it difficult to focus. You have hopefully spent lots of time preparing for the exam. Don't let an empty stomach get in the way of success!

When travelling to the testing center, leave earlier than needed. That way, you have a buffer in case you experience any delays. This will help you remain calm and will keep you from missing your appointment time at the testing center.

Be sure to pace yourself during the exam. Don't try to rush through the exam. There is no need to risk performing poorly on the exam just so you can leave the testing center early. Allow yourself to use all of the allotted time if needed.

Remain positive while taking the exam even if you feel like you are performing poorly. Thinking about the content you should have mastered will not help you perform better on the exam.

Once the exam is complete, take some time to relax. Even if you feel that you need to take the exam again, you will be well served by some down time before you begin studying again. It's often easier to convince yourself to study if you know that it will come with a reward!

Test-Taking Strategies

1. Predicting the Answer

When you feel confident in your preparation for a multiple-choice test, try predicting the answer before reading the answer choices. This is especially useful on questions that test objective factual knowledge or that ask you to fill in a blank. By predicting the answer before reading the available choices, you eliminate the possibility that you will be distracted or led astray by an incorrect answer choice. You will feel much more confident in your selection if you read the question, predict the answer, and then find your prediction among the answer choices. After using this strategy, be sure to still read all of the answer choices carefully and completely. If you feel unprepared, you should not attempt to predict the answers. This would be a waste of time and an opportunity for your mind to wander in the wrong direction.

2. Reading the Whole Question

Too often, test takers scan a multiple-choice question, recognize a few familiar words, and immediately jump to the answer choices. Test authors are aware of this common impatience, and they will sometimes prey upon it. For instance, a test author might subtly turn the question into a negative, or he or she might redirect the focus of the question right at the end. The only way to avoid falling into these traps is to read the entirety of the question carefully before reading the answer choices.

3. Looking for Wrong Answers

Long and complicated multiple-choice questions can be intimidating. One way to simplify a difficult multiple-choice question is to eliminate all of the answer choices that are clearly wrong. In most sets of answers, there will be at least one selection that can be dismissed right away. If the test is administered on paper, the test taker could draw a line through it to indicate that it may be ignored; otherwise, the test taker will have to perform this operation mentally or on scratch paper. In either case, once the obviously incorrect answers have been eliminated, the remaining choices may be considered. Sometimes identifying the clearly wrong answers will give the test taker some information about the correct answer. For instance, if one of the remaining answer choices is a direct opposite of one of the eliminated answer choices, it may well be the correct answer. The opposite of obviously wrong is obviously right! Of course, this is not always the case. Some answers are obviously incorrect simply because they are irrelevant to the question being asked. Still, identifying and eliminating some incorrect answer choices is a good way to simplify a multiple-choice question.

4. Don't Overanalyze

Anxious test takers often overanalyze questions. When you are nervous, your brain will often run wild causing you to make associations and discover clues that don't actually exist. If you feel that this may be a problem for you, do whatever you can to slow down during the test. Try taking a deep breath or counting to ten. As you read and consider the question, restrict yourself to the particular words used by the author. Avoid thought tangents about what the author *really* meant, or what he or she was *trying* to say. The only things that matter on a multiple-choice test are the words that are actually in the question. You must avoid reading too much into a multiple-choice question, or supposing that the writer meant something other than what he or she wrote.

5. No Need for Panic

It is wise to learn as many strategies as possible before taking a multiple-choice test, but it is likely that you will come across a few questions for which you simply don't know the answer. In this situation, avoid panicking. Because most multiple-choice tests include dozens of questions, the relative value of a single wrong answer is small. Moreover, your failure on one question has no effect on your success elsewhere on the test. As much as possible, you should compartmentalize each question on a multiple-choice test. In other words, you should not allow your feelings about one question to affect your success on the others. When you find a question that you either don't understand or don't know how to answer, just take a deep breath and do your best. Read the entire question slowly and carefully. Try rephrasing the question a couple of different ways. Then, read all of the answer choices carefully. After eliminating obviously wrong answers, make a selection and move on to the next question.

6. Confusing Answer Choices

When working on a difficult multiple-choice question, there may be a tendency to focus on the answer choices that are the easiest to understand. Many people, whether consciously or not, gravitate to the answer choices that require the least concentration, knowledge, and memory. This is a mistake. When you come across an answer choice that is confusing, you need to give it extra attention. A question might be confusing because you do not know the subject matter to which it refers. If this is the case, don't eliminate the answer before you have affirmatively settled on another. When you come across an answer choice of this type, set it aside as you look at the remaining choices. If you can confidently assert that one of the other choices is correct, you can leave the confusing answer aside. Otherwise, you will need to take a moment to try to better understand the confusing answer choice. Rephrasing is one way to tease out the sense of a confusing answer choice.

7. Your First Instinct

Many people struggle with multiple-choice tests because they overthink the questions. If you have studied sufficiently for the test, you should be prepared to trust your first instinct once you have carefully and completely read the question and all of the answer choices. There is a great deal of research suggesting that the mind can come to the correct conclusion very quickly once it has obtained all of the relevant information. At times, it may seem to you as if your intuition is working faster even than your reasoning mind. This may in fact be true. The knowledge you obtain while studying may be retrieved from your subconscious before you have a chance to work out the associations that support it. Verify your instinct by working out the reasons that it should be trusted.

8. Key Words

Many test takers struggle with multiple-choice questions because they have poor reading comprehension skills. Quickly reading and understanding a multiple-choice question requires a mixture of skill and experience. To help with this, try jotting down a few key words and phrases on a piece of scrap paper. Doing this concentrates the process of reading and forces the mind to weigh the relative importance of the question's parts. In selecting words and phrases to write down, the test taker thinks about the question more deeply and carefully. This is especially true for multiple-choice questions that are preceded by a long prompt.

9. Subtle Negatives

One of the oldest tricks in the multiple-choice test writer's book is to subtly reverse the meaning of a question with a word like *not* or *except*. If you are not paying attention to each word in the question, you can easily be led astray by this trick. For instance, a common question format is, "Which of the following is...?" Obviously, if the question instead is, "Which of the following is not....?," then the answer will be quite different. Even worse, the test makers are aware of the potential for this mistake and will include one answer choice that would be correct if the question were not negated or reversed. A test taker who misses the reversal will find what he or she believes to be a correct answer and will be so confident that he or she will fail to reread the question and discover the original error. The only way to avoid this is to practice a wide variety of multiple-choice questions and to pay close attention to each and every word.

10. Reading Every Answer Choice

It may seem obvious, but you should always read every one of the answer choices! Too many test takers fall into the habit of scanning the question and assuming that they understand the question because they recognize a few key words. From there, they pick the first answer choice that answers the question they believe they have read. Test takers who read all of the answer choices might discover that one of the latter answer choices is actually *more* correct. Moreover, reading all of the answer choices can remind you of facts related to the question that can help you arrive at the correct answer. Sometimes, a misstatement or incorrect detail in one of the latter answer choices will trigger your memory of the subject and will enable you to find the right answer. Failing to read all of the answer choices is like not reading all of the items on a restaurant menu: you might miss out on the perfect choice.

11. Spot the Hedges

One of the keys to success on multiple-choice tests is paying close attention to every word. This is never more true than with words like *almost*, *most*, *some*, and *sometimes*. These words are called "hedges", because they indicate that a statement is not totally true or not true in every place and time. An absolute statement will contain no hedges, but in many subjects, like literature and history, the answers are not always straightforward or absolute. There are always exceptions to the rules in these subjects. For this reason, you should favor those multiple-choice questions that contain hedging language. The presence of qualifying words indicates that the author is taking special care with his or her words, which is certainly important when composing the right answer. After all, there are many ways to be wrong, but there is only one way to be right! For this reason, it is wise to avoid answers that are absolute when taking a multiple-choice test. An absolute answer is one that says things are either all one way or all another. They often include words like *every*, *always*, *best*, and *never*. If you are taking a multiple-choice test in a subject that doesn't lend itself to absolute answers, be on your guard if you see any of these words.

12. Long Answers

In many subject areas, the answers are not simple. As already mentioned, the right answer often requires hedges. Another common feature of the answers to a complex or subjective question are qualifying clauses, which are groups of words that subtly modify the meaning of the sentence. If the question or answer choice describes a rule to which there are exceptions or the subject matter is complicated, ambiguous, or confusing, the correct answer will require many words in order to be expressed clearly and accurately. In essence, you should not be deterred by answer choices that seem excessively long. Oftentimes, the author of the text will not be able to write the correct answer without offering some qualifications and modifications. As a test taker, your job is to read the answer choices thoroughly and completely and to select the one that most accurately and precisely answers the question.

13. Restating to Understand

Sometimes, a question on a multiple-choice test is difficult not because of what it asks but because of how it is written. If this is the case, restate the question or answer choice in different words. This process serves a couple of important purposes. First, it forces you to concentrate on the core of the question. In order to rephrase the question accurately, you have to understand it well. Rephrasing the question will concentrate your mind on the key words and ideas. Second, it will present the information to your mind in a fresh way. This process may trigger your memory and render some useful scrap of information picked up while studying.

14. True Statements

Sometimes an answer choice will be true in itself, but it does not answer the question. This is one of the main reasons why it is essential to read the question carefully and completely before proceeding to the answer choices. Too often, test takers skip ahead to the answer choices and look for true statements. Having found one of these, they are content to select it without reference to the question above. Obviously, this provides an easy way for test makers to play tricks. The savvy test taker will always read the entire question before turning to the answer choices. Then, having settled on a correct answer choice, he or she will refer to the original question and ensure that the selected answer is relevant. The mistake of choosing a correct-but-irrelevant answer choice is especially common on questions related to specific pieces of objective knowledge, like historical or scientific facts. A prepared test taker will have a wealth of factual knowledge at his or her disposal, and should not be careless in its application.

15. No Patterns

One of the more dangerous ideas that circulates about multiple-choice tests is that the correct answers tend to fall into patterns. These erroneous ideas range from a belief that B and C are the most common right answers, to the idea that an unprepared test-taker should answer "A-B-A-C-A-D-A-B-A." It cannot be emphasized enough that pattern-seeking of this type is exactly the WRONG way to approach a multiple-choice test. To begin with, it is highly unlikely that the test maker will plot the correct answers according to some predetermined pattern. The questions are scrambled and delivered in a random order. Furthermore, even if the test maker was following a pattern in the assignation of correct answers, there is no reason why the test taker would know which pattern he or she was using. Any attempt to discern a pattern in the answer choices is a waste of time and a distraction from the real work of taking the test. A test taker would be much better served by extra preparation before the test than by reliance on a pattern in the answers.

FREE DVD OFFER

Don't forget that doing well on your exam includes both understanding the test content and understanding how to use what you know to do well on the test. We offer a completely FREE Test Taking Tips DVD that covers world class test taking tips that you can use to be even more successful when you are taking your test.

All that we ask is that you email us your feedback about your study guide. To get your **FREE Test Taking Tips DVD**, email freedvd@studyguideteam.com with "FREE DVD" in the subject line and the following information in the body of the email:

- The title of your study guide.
- Your product rating on a scale of 1-5, with 5 being the highest rating.
- Your feedback about the study guide. What did you think of it?
- Your full name and shipping address to send your free DVD.

Introduction to the AFOQT

Function of the Test

The Air Force Officer Qualifying Test (AFOQT) is a standardized test given by the United States Air Force. The exam evaluates a test taker's verbal and mathematical proficiency as well as his or her aptitude in certain areas specific to those necessary for potential Air Force career paths. The test is used as part of the admissions process to officer training programs, such as Officer Training School ROTC. Within the Air Force, it is used to qualify candidates for Pilot, Combat Systems Officer (CSO), and Air Battle Manager (ABM) training and is part of the Pilot Candidate Selection Method (PCSM) score. The AFOQT is required for all students receiving a scholarship as well as those in the Professional Officer Course (POC).

The test is taken nationwide by current and potential members of the United States Air Force. In the Air Force ROTC program, it is taken by sophomores prior to field training in the summer after their sophomore year.

Test Administration

The AFOQT is offered through Air Force ROTC programs on college campuses and through military recruiters at Military Entrance Processing facilities. There is no cost to take the AFOQT; instead, individuals wishing to take the test must make arrangements through their ROTC program, recruiter, or commanding officer, as appropriate. Rules for retesting depend on the purpose or program for which the test taker is seeking to use the results, but some ROTC programs permit one retest, with the most recent score counting.

Test Format

The test lasts almost five hours, including three hours and 36.5 minutes of testing time and a little over an hour in breaks and test administration time. It is taken with pencil and scored by machine. It consists of twelve subtests: verbal analogies, arithmetic reasoning, word knowledge, math knowledge, reading comprehension, situational judgment, self-description inventory, physical science, table reading, instrument comprehension, block counting, and aviation information. All of the subtests have multiple-choice questions with four or five possible answers.

Scoring

Scores are based only on the number of correct answers. There is no penalty for guessing incorrectly, aside from the missed opportunity to achieve points from a greater number of correct answers. Scores from the various subtests are used to calculate composite scores, which are reported to the test taker and the Air Force. For example, the "Pilot" composite score is based on the results from the arithmetic reasoning, math knowledge, instrument comprehension, table reading, and aviation information subtests. Other composite scores include Academic Aptitude, Verbal, Quantitative, Combat Systems Officer, Air Battle Manager, and Situational Judgment. Test takers receive a percentile score from 1 to 99 in each of the five composite categories.

There is no set passing score. Instead, the scores needed vary widely depending on the intended job or program for which a test taker is seeking entry. For instance, a candidate seeking to become an officer may be able to do so with a relatively low percentile score (in other words, by only outperforming a small number of other test takers), while an officer seeking to become a pilot may need much higher scores overall, particularly in the Pilot composite category.

Recent/Future Developments

The AFOQT is revised from time to time, based on feedback from the general needs of the Air Force and its officer training programs. The current subtests and content therein are in AFOQT Form T, which took effect on August 1, 2014.

A summary of the number of items on and the time allowed for each subtest is as follows:

Subtest	Items	Time (min.)
Verbal Analogies	25	8
Arithmetic Reasoning	25	29
Word Knowledge	25	5
Math Knowledge	25	22
Self-Description Inventory	220	40
General Science	20	10
Table Reading	40	7
Instrument Comprehension	20	6
Block Counting	20	3
Aviation Information	20	8
Rotated Blocks	15	13
Hidden Figures	15	8
TOTAL	**470**	**3.5 hours**

Verbal Analogies

Verbal Analogies

The verbal analogies test portion of the AFOQT tests the candidate's ability to analyze words carefully and find connections in definition and/or context. The test-taker must compare a selected set of words with answer choices and select the ideal word to complete the sequence. While these exercises draw upon knowledge of vocabulary, this is also a test of critical thinking and reasoning abilities. Naturally, such skills are critical for building a career. Mastering verbal analogies will help people think objectively, discern critical details, and communicate more efficiently.

Question Layout

Verbal analogy sections are on other standardized tests such as the SAT. The format on the AFOQT remains basically the same. First, two words are paired together that provide a frame for the analogy, and then there is a third word that must be found as a match in kind. It may help to think of it like this: A is to B as C is to D. Examine the breakdown below:

Apple (A) is to fruit (B) as carrot (C) is to vegetable (D).

As shown above, there are four words: the first three are given and the fourth word is the answer that must be found. The first two words are given to set up the kind of analogy that is to be replicated for the next pair. We see that apple is paired with fruit. In the first pair, a specific food item, apple, is paired to the food group category it corresponds with, which is fruit. When presented with the third word in the verbal analogy, carrot, a word must be found that best matches carrot in the way that fruit matched with apple. Again, carrot is a specific food item, so a match should be found with the appropriate food group: vegetable! Here's a sample prompt:

Morbid is to dead as jovial is to
- a. Hate.
- b. Fear.
- c. Disgust.
- d. Happiness.
- e. Desperation.

As with the apple and carrot example, here is an analogy frame in the first two words: morbid and dead. Again, this will dictate how the next two words will correlate with one another. The definition of morbid is: described as or appealing to an abnormal and unhealthy interest in disturbing and unpleasant subjects, particularly death and disease. In other words, morbid can mean ghastly or death-like, which is why the word dead is paired with it. Dead relates to morbid because it describes morbid. With this in mind, jovial becomes the focus. Jovial means joyful, so out of all the choices given, the closest answer describing jovial is happiness (D).

Prompts on the exam will be structured just like the one above. "A is to B as C is to ?" will be given, where the answer completes the second pair. Or sometimes, "A is to B as ? is to ?" is given, where the second pair of words must be found that replicate the relationship between the first pair. The only things that will change are the words and the relationships between the words provided.

Discerning the Correct Answer

While it wouldn't hurt in test preparation to expand vocabulary, verbal analogies are all about delving into the words themselves and finding the right connection, the right word that will fit an analogy. People preparing for the test shouldn't think of themselves as human dictionaries, but rather as detectives. Remember, how the first two words are connected dictates the second pair. From there, picking the correct answer or simply eliminating the ones that aren't correct is the best strategy.

Just like a detective, a test-taker needs to carefully examine the first two words of the analogy for clues. It's good to get in the habit of asking the questions: What do the two words have in common? What makes them related or unrelated? How can a similar relationship be replicated with the word I'm given and the answer choices? Here's another example:

Pillage is to steal as meander is to
    a. Stroll.
    b. Burgle.
    c. Cascade.
    d. Accelerate.
    e. Pinnacle.

Why is pillage paired with steal? In this example, pillage and steal are synonymous: they both refer to the act of stealing. This means that the answer is a word that means the same as meander, which is stroll. In this case, the defining relationship in the whole analogy was a similar definition.

What if test-takers don't know what stroll or meander mean, though? Using logic helps to eliminate choices and pick the correct answer. Looking closer into the contexts of the words pillage and steal, here are a few facts: these are things that humans do; and while they are actions, these are not necessarily types of movement. Again, pick a word that will not only match the given word, but best completes the relationship. It wouldn't make sense that burgle (B) would be the correct choice because meander doesn't have anything to do with stealing, so that eliminates burgle. Pinnacle (E) also can be eliminated because this is not an action at all but a position or point of reference. Cascade (C) refers to pouring or falling, usually in the context of a waterfall and not in reference to people, which means we can eliminate cascade as well. While people do accelerate when they move, they usually do so under additional circumstances: they accelerate while running or driving a car. All three of the words we see in the analogy are actions that can be done independently of other factors. Therefore, accelerate (D) can be eliminated, and stroll (A) should be chosen. Stroll and meander both refer to walking or wandering, so this fits perfectly.

The *process of elimination* will help rule out wrong answers. However, the best way to find the correct answer is simply to differentiate the correct answer from the other choices. For this, test-takers should go back to asking questions, starting with the chief question: What's the connection? There are actually many ways that connections can be found between words. The trick is to look for the answer that is consistent with the relationship between the words given. What is the prevailing connection? Here are a few different ways verbal analogies can be formed.

Finding Connections in Word Analogies

<u>Connections in Categories</u>
One of the easiest ways to choose the correct answer in word analogies is simply to group words together. Ask if the words can be compartmentalized into *distinct categories*. Here are some examples:

Terrier is to dog as mystery is to
- a. Thriller.
- b. Murder.
- c. Detective.
- d. Novel.
- e. Investigation.

This one might have been a little confusing, but when looking at the first two words in the analogy, this is clearly one in which a category is the prevailing theme. Think about it: a terrier is a type of dog. While there are several breeds of dogs that can be categorized as a terrier, in the end, all terriers are still dogs. Therefore, mystery needs to be grouped into a category. Murders, detectives, and investigations can all be involved in a mystery plot, but a murder (B), a detective (C), or an investigation (E) is not necessarily a mystery. A thriller (A) is a purely fictional concept, a kind of story or film, just like a mystery. A thriller can describe a mystery, but the same issue appears as the other choices. What about novel (D)? For one thing, it's distinct from all the other terms. A novel isn't a component of a mystery, but a mystery can be a type of novel. The relationship fits: a terrier is a type of dog, just like a mystery is a type of novel.

<u>Synonym/Antonym</u>
Some analogies are based on words meaning the same thing or expressing the same idea. Sometimes it's the complete opposite!

Marauder is to brigand as
- a. King is to peasant.
- b. Juice is to orange.
- c. Soldier is to warrior.
- d. Engine is to engineer.
- e. Paper is to photocopier.

Here, soldier is to warrior (C) is the correct answer. Marauders and brigands are both thieves. They are synonyms. The only pair of words that fits this analogy is soldier and warrior because both terms describe combatants who fight.

Cap is to shoe as jacket is to
- a. Ring.
- b. T-shirt.
- c. Vest.
- d. Glasses.
- e. Pants.

Opposites are at play here because a cap is worn on the head/top of the person, while a shoe is worn on the foot/bottom. A jacket is worn on top of the body too, so the opposite of jacket would be pants (E) because these are worn on the bottom of the body. Often the prompts on the test provide a synonym or antonym relationship. Just consider if the sets in the prompt reflect similarity or stark difference.

Parts of a Whole

Another thing to consider when first looking at an analogy prompt is whether the words presented come together in some way. Do they express parts of the same item? Does one word complete the other? Are they connected by process or function?

Tire is to car as
    a. Wing is to bird.
    b. Oar is to boat.
    c. Box is to shelf.
    d. Hat is to head.
    e. Knife is to sheath.

We know that the tire fits onto the car's wheels and this is what enables the car to drive on roads. The tire is part of the car. This is the same relationship as oar is to boat (B). The oars are attached onto a boat and enable a person to move and navigate the boat on water. At first glance, wing is to bird (A) seems to fit too, since a wing is a part of a bird that enables it to move through the air. However, since a tire and car are not alive and transport people, oar and boat fit better because they are also not alive and they transport people. Subtle differences between answer choices should be found.

Other Relationships

There are a number of other relationships to look for when solving verbal analogies. Some relationships focus on one word being a *characteristic/NOT a characteristic* of the other word. Sometimes the first word is *the source/comprised of* the second word. Still, other words are related by their *location*. Some analogies have *sequential* relationships, and some are *cause/effect* relationships. There are analogies that show *creator/provider* relationships with the *creation/provision*. Another relationship might compare an *object* with its *function* or a *user* with his or her *tool*. An analogy may focus on a *change of grammar* or a *translation of language*. Finally, one word of an analogy may have a relationship to the other word in its *intensity*. The type of relationship between the first two words of the analogy should be determined before continuing to analyze the second set of words. One effective method of determining a relationship between two words is to form a comprehensible sentence using both words and then to plug the answer choices into the same sentence. For example, consider the following analogy: *Bicycle is to handlebars as car is to steering wheel*. A sentence could be formed that says: A bicycle navigates using its handlebars; therefore, a car navigates using its steering wheel. If the second sentence makes sense, then the correct relationship has likely been found. A sentence may be more complex depending on the relationship between the first two words in the analogy. An example of this may be: *food is to dishwasher as dirt is to carwash*. The formed sentence may be: A dishwasher cleans food off of dishes in the same way that a carwash cleans dirt off of a car.

Dealing with Multiple Connections

There are many other ways to draw connections between word sets. Several word choices might form an analogy that would fit the word set in your prompt. When this occurs, the analogy must be explored from multiple angles as, on occasion, multiple answer choices may appear to be correct. When this occurs, ask yourself: which one is an even closer match than the others? The framing word pair is another important point to consider. Can one or both words be interpreted as actions or ideas, or are they purely objects? Here's a question where words could have alternate meanings:

Hammer is to nail as saw is to
    a. Electric.
    b. Hack.
    c. Cut.
    d. Machete.
    e. Groove.

Looking at the question above, it becomes clear that the topic of the analogy involves construction tools. Hammers and nails are used in concert, since the hammer is used to pound the nail. The logical first thing to do is to look for an object that a saw would be used on. Seeing that there is no such object among the answer choices, a test-taker might begin to worry. After all, that seems to be the choice that would complete the analogy—but that doesn't mean it's the only choice that may fit. Encountering questions like this tests one's ability to see multiple connections between words—don't get stuck thinking that words can only be connected in a single way. The first two words given can be verbs instead of just objects. To hammer means to hit or beat; oftentimes it refers to beating something into place. This is also what nail means when it is used as a verb. Here are the word choices that reveal the answer.

First, it's known that a saw, well, saws. It uses a steady motion to cut an object, and indeed to saw means to cut! Cut (C) is one of our answer choices, but the other options should be reviewed. While some tools are electric (a), the use of power in the tools listed in the analogy isn't a factor. Again, it's been established that these word choices are not tools in this context. Therefore, machete (D) is also ruled out because machete is also not a verb. Another important thing to consider is that while a machete is a tool that accomplishes a similar purpose as a saw, the machete is used in a slicing motion rather than a sawing/cutting motion. The verb that describes machete is hack (B), another choice that can be ruled out. A machete is used to hack at foliage. However, a saw does not hack. Groove (E) is just a term that has nothing to do with the other words, so this choice can be eliminated easily. This leaves cut (C), which confirms that this is the word needed to complete the analogy.

Practice Questions

1. **Cat** is to **paws** as
   a. Giraffe is to neck.
   b. Elephant is to ears.
   c. Horse is to hooves.
   d. Snake is to skin.
   e. Turtle is to shell.

2. **Dancing** is to **rhythm** as **singing** is to
   a. Pitch.
   b. Mouth.
   c. Sound.
   d. Volume.
   e. Words.

3. **Towel** is to **dry** as **hat** is to
   a. Cold.
   b. Warm.
   c. Expose.
   d. Cover.
   e. Top.

4. **Sand** is to **glass** as
   a. Protons are to atoms.
   b. Ice is to snow.
   c. Seeds are to plants.
   d. Water is to steam.
   e. Air is to wind.

5. **Design** is to **create** as **allocate** is to
   a. Finish.
   b. Manage.
   c. Multiply.
   d. Find.
   e. Distribute.

6. **Books** are to **reading** as
   a. Movies are to making.
   b. Shows are to watching.
   c. Poetry is to writing.
   d. Scenes are to performing.
   e. Concerts are to music.

7. **Cool** is to **frigid** as **warm** is to
   a. Toasty.
   b. Summer.
   c. Sweltering.
   d. Hot.
   e. Mild.

8. **Buses** are to **rectangular prisms** as **globes** are to
   a. Circles.
   b. Maps.
   c. Wheels.
   d. Spheres.
   e. Movement.

9. **Backpacks** are to **textbooks** as
   a. Houses are to people.
   b. Fences are to trees.
   c. Plates are to food.
   d. Chalkboards are to chalk.
   e. Computers are to mice.

10. **Storm** is to **rainbow** as **sunset** is to
   a. Clouds.
   b. Sunrise.
   c. Breakfast.
   d. Bedtime.
   e. Stars.

11. **Falcon** is to **mice** as **giraffe** is to
   a. Leaves.
   b. Rocks.
   c. Antelope.
   d. Grasslands.
   e. Hamsters.

12. **Car** is to **motorcycle** as **speedboat** is to
   a. Raft.
   b. Jet-ski.
   c. Sailboat.
   d. Plane.
   e. Canoe.

13. **Arid** is to **damp** as **anxious** is to
   a. Happy.
   b. Petrified.
   c. Ireful.
   d. Confident.
   e. Sorrowful.

14. **Mechanic** is to **repair** as
    a. Mongoose is to cobra.
    b. Rider is to bicycle.
    c. Tree is to grow.
    d. Food is to eaten.
    e. Doctor is to heal.

15. **Whistle** is to **blow horn** as **painting** is to
    a. View.
    b. Criticize.
    c. Sculpture.
    d. Painter.
    e. Paintbrush.

16. **Paddle** is to **boat** as **keys** are to
    a. Unlock.
    b. Success.
    c. Illuminate.
    d. Piano.
    e. Keychain.

17. **Monotonous** is to **innovative** as
    a. Gorgeous is to beautiful.
    b. Ancient is to archaic.
    c. Loquacious is to silent.
    d. Sturdy is to fortified.
    e. Spectacular is to grandiose.

18. **Mountain** is to **peak** as **wave** is to
    a. Ocean.
    b. Surf.
    c. Fountain.
    d. Wavelength.
    e. Crest.

19. **Ambiguous** is to **indecisive** as **uncertain** is to
    a. Indefinite.
    b. Certain.
    c. Flippant.
    d. Fearful.
    e. Rounded.

20. **Fluent** is to **communication** as
    a. Crater is to catastrophe.
    b. Gourmet is to cooking.
    c. Ink is to pen.
    d. Crow is to raven.
    e. Whistle is to whistler.

21. **Validate** is to **truth** as **conquer** is to
    a. Withdraw.
    b. Subjugate.
    c. Expand.
    d. Surrender.
    e. Expose.

22. **Winter** is to **autumn** as **summer** is to
    a. Vacation.
    b. Fall.
    c. Spring.
    d. March.
    e. Weather.

23. **Penguin** is to **lemur** as **eagle** is to
    a. Howler monkey.
    b. Osprey.
    c. Warthog.
    d. Kestrel.
    e. Moose.

24. **Fiberglass** is to **surfboard** as
    a. Bamboo is to panda.
    b. Capital is to D.C.
    c. Copper is to penny.
    d. Flint is to mapping.
    e. Wind is to windmill.

25. **Myth** is to **explain** as **joke** is to
    a. Enlighten.
    b. Inspire.
    c. Collect.
    d. Laughter.
    e. Amuse.

23

Answer Explanations

1. C: This is a part/whole analogy. The common thread is what animals walk on. Choices *A*, *B*, and *E* all describe signature parts of animals, but paws are not the defining feature of cats. While snakes travel on their skins, they do not walk.

2. A: This is a characteristic analogy. The connection lies in what observers will judge a performance on. While the other choices are also important, an off-key singer is as unpleasant as a dancer with no rhythm.

3. D: This is a use/tool analogy. The analogy focuses on an item's use. While hats are worn when it's cold with the goal of making the top of your head warm, this is not always guaranteed—their primary use is to provide cover. There is also the fact that not all hats are used to keep warm, but all hats cover the head.

4. D: This is a source/comprised of analogy. The common thread is addition of fire. Protons contribute to atoms and seeds grow into plants, but these are simple matters of building and growing without necessarily involving fire. B and E relate objects that already have similar properties.

5. E: This is a synonym analogy. The determining factor is synonymous definition. Design and create are synonyms, as are allocate and distribute. Typically, items are found and allocated as part of management to finish a project, but these qualities are not innate in the word. Allocation generally refers to the division of commodities instead of multiplication.

6. B: This is a tool/use analogy. The common thread is audience response to an art form. *A*, *C*, and *D* deal with the creation of artwork instead of its consumption. Choice *E* describes a form of art instead of the audience engagement with such.

7. C: This is an intensity analogy. The common thread is degree of severity. While *A*, *D*, and *E* can all describe warmth, they don't convey the harshness of sweltering. *B* simply describes a time when people may be more likely to think of warmth.

8. D: This is a characteristic analogy and is based on matching objects to their geometric shapes. Choice *A* is not correct because globes are three-dimensional, whereas circles exist in two dimensions. While wheels are three-dimensional, they are not always solid or perfectly round.

9. A: This is a tool/use analogy. The key detail of this analogy is the idea of enclosing or sealing items/people. When plates are filled with food, there is no way to enclose the item. While trees can be inside a fence, they can also be specifically outside of one.

10. E: This is a sequence of events analogy. The common thread is celestial cause-and-effect. Not everyone has breakfast or goes to bed after sunset. Sunrise is not typically thought of as the next interesting celestial event after sunsets. While clouds can develop after sunsets, they are also present before and during this activity. Stars, however, can be seen after dark.

11. A: This is a provider/provision analogy. The theme of this analogy is pairing a specific animal to their food source. Falcons prey on mice. Giraffes are herbivores and only eat one of the choices: leaves. Grasslands describe a type of landscape, not a food source for animals.

12. B: This is a category analogy. The common thread is motorized vehicles. While Choices *A*, *C*, and *E* also describe vehicles that move on water, they are not motorized. Although relying on engines, planes are not a form of water transportation.

13. D: This is an antonym analogy. The prevailing connection is opposite meanings. While happy can be an opposite of anxious, it's also possible for someone to experience both emotions at once. *B*, *C*, and *E* are also concurrent with anxious, not opposite.

14. E: This is a provider/provision analogy. This analogy looks at professionals and what their job is. Just as a mechanic's job is to repair machinery, a doctor works to heal patients.

15. C: This is a category analogy. Both whistles and blow horns are devices used to project/produce sound. Therefore, the analogy is based on finding something of a categorical nature. While *A*, *B*, *D*, and *E* involve or describe painting, they do not pertain to a distinct discipline alongside painting. Sculpture, however, is another form of art and expression, just like painting.

16. D: This is a part/whole analogy. This analogy examines the relationship between two objects. Specifically, this analogy examines how one object connects to another object, with the first object(s) being the means by which people use the corresponding object to produce a result directly. People use a paddle to steer a boat, just as pressing keys on a piano produces music. *B* and *C* can be metaphorically linked to keys but are unrelated. *A* is related to keys but is a verb, not another object. Choice *E* is the trickiest alternative, but what's important to remember is that while keys are connected to key chains, there is no result just by having the key on a key chain.

17. C: This is an antonym analogy. The common thread is opposite meanings. Monotonous refers to being dull or being repetitive, while innovative means new and bringing in changes. All of the other choices reflect synonymous word pairs. However, loquacious, which means talkative, is the opposite of silent.

18. E: This is a part/whole analogy. This analogy focuses on natural formations and their highest points. The peak of a mountain is its highest point just as the crest is the highest rise in a wave.

19. A: This is a negative connotations analogy. Essentially all of the given words in the example express the same idea of uncertainty or not taking a definitive stance. Ambiguous means open to possibilities, which parallels to not being able to make a decision, which describes indecisive as well as uncertain. Uncertain also means not definite, which not only relates to the given words but also directly to indefinite.

20. B: This is an intensity analogy. Fluent refers to how well one can communicate, while gourmet describes a standard of cooking. The analogy draws on degrees of a concept.

21. B: This is a synonym analogy, which relies on matching terms that are most closely connected. Validate refers to finding truth. Therefore, finding the term that best fits conquer is a good strategy. While nations have conquered others to expand their territory, they are ultimately subjugating those lands and people to their will. Therefore, subjugate is the best-fitting answer.

22. C: This is a sequence of events analogy. This analogy pairs one season with the season that precedes it. Winter is paired with autumn because autumn actually comes before winter. Out of all the answers, only *B* and *C* are actual seasons. Fall is another name for autumn, which comes after summer, not before. Spring, of course, is the season that comes before summer, making it the right answer.

23. A: This is a pairs analogy. None of the given terms are really related. To find the analogy, the way that each term is paired should be analyzed. Penguin, a bird, is paired with lemur, which is a primate. When given eagle, the only logical analogy to be made is to find another primate to pair with a bird, which is howler monkey.

24. C: This is a source/comprised of analogy. This analogy focuses on pairing a raw material with an object that it's used to create. Fiberglass is used to build surfboards just as copper is used in the creation of pennies. While wind powers a windmill, there is no physical object produced, like with the fiberglass/surfboard pair.

25. E: This is an object/function analogy. The common thread between these words is that one word describes a kind of story and it is paired with the purpose of the story. Myth is/was told in order to explain fundamental beliefs and natural phenomena. While laughter can result from a joke, the purpose of telling a joke is to amuse the audience, thus making *E* the right choice.

Arithmetic Reasoning

The Scope of the Arithmetic Reasoning Section

Problems in the Arithmetic Reasoning section of the AFOQT are generally word problems, which will require the use of reasoning and mathematics to find a solution. The problems normally present some everyday situations, along with a list of choices for answers. Some of the things to know include rates, speeds, percentages, averages, fractions, and ratios. The practice problems given later will cover the different types of questions in this section, although every word problem is slightly different.

How to Prepare

These problems involve basic arithmetic skills as well as the ability to break down a word problem to see where to apply these skills in order to get the correct answer. The basics of arithmetic and the approach to solving word problems are discussed here.

Note that math requires practice in order to become proficient. Make sure to not just read through the material here, but also try out the practice questions, as well as check the answers provided. Just reading through examples does not necessarily mean that a student can do the problems themselves. Note that sometimes there can be multiple approaches to getting a solution when doing the problems. What matters is getting the correct answer, so it is okay if the approach to a problem is different than the solution method provided.

Basic Operations of Arithmetic

There are four different basic operations used with numbers: addition, subtraction, multiplication, and division.

- Addition takes two numbers and combines them into a total called the sum. The sum is the total when combining two collections into one. If there are 5 things in one collection and 3 in another, then after combining them, there is a total of $5 + 3 = 8$. Note the order does not matter when adding numbers. For example, $3 + 5 = 8$.

- Subtraction is the opposite (or "inverse") operation to addition. Whereas addition combines two quantities together, subtraction takes one quantity away from another. For example, if there are 20 gallons of fuel and 5 are removed, that gives $20 - 5 = 15$ gallons remaining. Note that for subtraction, the order does matter because it makes a difference which quantity is being removed from which.

- Multiplication is repeated addition. 3×4 can be thought of as putting together 3 sets of items, each set containing 4 items. The total is 12 items. Another way to think of this is to think of each number as the length of one side of a rectangle. If a rectangle is covered in tiles with 3 columns of 4 tiles each, then there are 12 tiles in total. From this, one can see that the answer is the same if the rectangle has 4 rows of 3 tiles each: $4 \times 3 = 12$. By expanding this reasoning, the order the numbers are multiplied does not matter.

- Division is the opposite of multiplication. It means taking one quantity and dividing it into sets the size of the second quantity. If there are 16 sandwiches to be distributed to 4 people, then each person gets $16 \div 4 = 4$ sandwiches. As with subtraction, the order in which the numbers appear does matter for division.

Addition

Addition is the combination of two numbers so their quantities are added together cumulatively. The sign for an addition operation is the + symbol. For example, 9 + 6 = 15. The 9 and 6 combine to achieve a cumulative value, called a sum.

Addition holds the commutative property, which means that numbers in an addition equation can be switched without altering the result. The formula for the commutative property is a + b = b + a. Let's look at a few examples to see how the commutative property works:

$$7 = 3 + 4 = 4 + 3 = 7$$

$$20 = 12 + 8 = 8 + 12 = 20$$

Addition also holds the associative property, which means that the grouping of numbers doesn't matter in an addition problem. In other words, the presence or absence of parentheses is irrelevant. The formula for the associative property is $(a + b) + c = a + (b + c)$. Here are some examples of the associative property at work:

$$30 = (6 + 14) + 10 = 6 + (14 + 10) = 30$$

$$35 = 8 + (2 + 25) = (8 + 2) + 25 = 35$$

Subtraction

Subtraction is taking away one number from another, so their quantities are reduced. The sign designating a subtraction operation is the − symbol, and the result is called the difference. For example, $9 - 6 = 3$. The number *6* detracts from the number *9* to reach the difference *3*.

Unlike addition, subtraction follows neither the commutative nor associative properties. The order and grouping in subtraction impact the result.

$$15 = 22 - 7 \neq 7 - 22 = -15$$

$$3 = (10 - 5) - 2 \neq 10 - (5 - 2) = 7$$

When working through subtraction problems involving larger numbers, it's necessary to regroup the numbers. Let's work through a practice problem using regrouping:

$$
\begin{array}{r}
3\ 2\ 5 \\
-\ \ 7\ 7 \\
\hline
\end{array}
$$

Here, it is clear that the ones and tens columns for 77 are greater than the ones and tens columns for 325. To subtract this number, borrow from the tens and hundreds columns. When borrowing from a column, subtracting 1 from the lender column will add 10 to the borrower column:

$$
\begin{array}{ccc}
3\text{-}1 & 10+2\text{-}1 & 10+5 \\
- & 7 & 7
\end{array}
\ =\
\begin{array}{ccc}
2 & 11 & 15 \\
- & 7 & 7 \\
\hline
2 & 4 & 8
\end{array}
$$

After ensuring that each digit in the top row is greater than the digit in the corresponding bottom row, subtraction can proceed as normal, and the answer is found to be 248.

Multiplication

Multiplication involves adding together multiple copies of a number. It is indicated by an $\times$ symbol or a number immediately outside of a parentheses, e.g. $5(8 - 2)$. The two numbers being multiplied together are called factors, and their result is called a product. For example, $9 \times 6 = 54$. This can be shown alternatively by expansion of either the 9 or the 6:

$$9 \times 6 = 9 + 9 + 9 + 9 + 9 + 9 = 54$$

$$9 \times 6 = 6 + 6 + 6 + 6 + 6 + 6 + 6 + 6 + 6 = 54$$

Like addition, multiplication holds the commutative and associative properties:

$$115 = 23 \times 5 = 5 \times 23 = 115$$

$$84 = 3 \times (7 \times 4) = (3 \times 7) \times 4 = 84$$

Multiplication also follows the distributive property, which allows the multiplication to be distributed through parentheses. The formula for distribution is $a \times (b + c) = ab + ac$. This is clear after the examples:

$$45 = 5 \times 9 = 5(3 + 6) = (5 \times 3) + (5 \times 6) = 15 + 30 = 45$$

$$20 = 4 \times 5 = 4(10 - 5) = (4 \times 10) - (4 \times 5) = 40 - 20 = 20$$

Multiplication becomes slightly more complicated when multiplying numbers with decimals. The easiest way to answer these problems is to ignore the decimals and multiply as if they were whole numbers. After multiplying the factors, place a decimal in the product. The placement of the decimal is determined by taking the cumulative number of decimal places in the factors.

For example:

$$
\begin{array}{r}
0.7 \\
\times\ 3 \\
\hline
2.1
\end{array}
\qquad
\begin{array}{r}
2.6 \\
\times\ 4.2 \\
\hline
10.92
\end{array}
\qquad
\begin{array}{r}
1.5 \\
\times\ 6.4 \\
\hline
9.60
\end{array}
$$

Let's tackle the first example. First, ignore the decimal and multiply the numbers as though they were whole numbers to arrive at a product: 21. Second, count the number of digits that follow a decimal (one). Finally, move the decimal place that many positions to the left, as the factors have only one decimal place. The second example works the same way, except that there are two total decimal places in the factors, so the product's decimal is moved two places over. In the third example, the decimal should be moved over two digits, but the digit zero is no longer needed, so it is erased and the final answer is 9.6.

Division

Division and multiplication are inverses of each other in the same way that addition and subtraction are opposites. The signs designating a division operation are the ÷ and / symbols. In division, the second number divides into the first.

The number before the division sign is called the dividend or, if expressed as a fraction, the numerator. For example, in $a \div b$, a is the dividend, while in $\frac{a}{b}$, a is the numerator.

The number after the division sign is called the divisor or, if expressed as a fraction, the denominator. For example, in $a \div b$, b is the divisor, while in $\frac{a}{b}$, b is the denominator.

Like subtraction, division doesn't follow the commutative property, as it matters which number comes before the division sign, and division doesn't follow the associative or distributive properties for the same reason. For example:

$$\frac{3}{2} = 9 \div 6 \neq 6 \div 9 = \frac{2}{3}$$

$$2 = 10 \div 5 = (30 \div 3) \div 5 \neq 30 \div (3 \div 5) = 30 \div \frac{3}{5} = 50$$

$$25 = 20 + 5 = (40 \div 2) + (40 \div 8) \neq 40 \div (2 + 8) = 40 \div 10 = 4$$

If a divisor doesn't divide into a dividend an integer number of times, whatever is left over is termed the remainder. The remainder can be further divided out into decimal form by using long division; however, this doesn't always give a quotient with a finite amount of decimal places, so the remainder can also be expressed as a fraction over the original divisor.

Division with decimals is similar to multiplication with decimals in that when dividing a decimal by a whole number, ignore the decimal and divide as if it were a whole number.

Upon finding the answer, or quotient, place the decimal at the decimal place equal to that in the dividend.

$$15.75 \div 3 = 5.25$$

When the divisor is a decimal number, multiply both the divisor and dividend by 10. Repeat this until the divisor is a whole number, then complete the division operation as described above.

$$17.5 \div 2.5 = 175 \div 25 = 7$$

Fractions

A *fraction* is a number used to express a ratio. It is written as a number x over a line with another number y underneath: $\frac{x}{y}$, and can be thought of as x out of y equal parts. The number on top (x) is called the *numerator*, and the number on the bottom is called the *denominator* (y). It is important to remember the only restriction is that the denominator is not allowed to be 0.

Imagine that an apple pie has been baked for a holiday party, and the full pie has eight slices. After the party, there are five slices left. How could the amount of the pie that remains be expressed as a fraction? The numerator is 5 since there are 5 pieces left, and the denominator is 8 since there were eight total slices in the whole pie. Thus, expressed as a fraction, the leftover pie totals $\frac{5}{8}$ of the original amount.

Another way of thinking about fractions is like this: $\frac{x}{y} = x \div y$.

Two fractions can sometimes equal the same number even when they look different. The value of a fraction will remain equal when multiplying both the numerator and the denominator by the same number. The value of the fraction does not change when dividing both the numerator and the denominator by the same number. For example, $\frac{4}{8} = \frac{2}{4} = \frac{1}{2}$. If two fractions look different, but are actually the same number, these are *equivalent fractions*.

A number that can divide evenly into a second number is called a *divisor* or *factor* of that second number; 3 is a divisor of 6, for example. If the numerator and denominator in a fraction have no common factors other than 1, the fraction is said to be *simplified*. $\frac{2}{4}$ is not simplified (since the numerator and denominator have a factor of 2 in common), but $\frac{1}{2}$ is simplified. Often, when solving a problem, the final answer generally requires us to simplify the fraction.

It is often useful when working with fractions to rewrite them so they have the same denominator. This process is called finding a *common denominator*. The common denominator of two fractions needs to be a number that is a multiple of both denominators. For example, given $\frac{1}{6}$ and $\frac{5}{8}$, a common denominator is $6 \times 8 = 48$. However, there are often smaller choices for the common denominator. The smallest number that is a multiple of two numbers is called the *least common multiple* of those numbers. For this example, use the numbers 6 and 8. The multiples of 6 are 6, 12, 18, 24... and the multiples of 8 are 8, 16, 24..., so the least common multiple is 24. The two fractions are rewritten as $\frac{4}{24}, \frac{15}{24}$.

If two fractions have a common denominator, then the numerators can be added or subtracted. For example, $\frac{4}{5} - \frac{3}{5} = \frac{4-3}{5} = \frac{1}{5}$. If the fractions are not given with the same denominator, a common denominator needs to be found before adding or subtracting them.

It is always possible to find a common denominator by multiplying the denominators by each other. However, when the denominators are large numbers, this method is unwieldy, especially if the answer must be provided in its simplest form. Thus, it's beneficial to find the least common denominator of the fractions—the least common denominator is incidentally also the least common multiple.

Once equivalent fractions have been found with common denominators, simply add or subtract the numerators to arrive at the answer:

$$1) \frac{1}{2} + \frac{3}{4} = \frac{2}{4} + \frac{3}{4} = \frac{5}{4}$$

$$2) \frac{3}{12} + \frac{11}{20} = \frac{15}{60} + \frac{33}{60} = \frac{48}{60} = \frac{4}{5}$$

$$3) \frac{7}{9} - \frac{4}{15} = \frac{35}{45} - \frac{12}{45} = \frac{23}{45}$$

$$4) \frac{5}{6} - \frac{7}{18} = \frac{15}{18} - \frac{7}{18} = \frac{8}{18} = \frac{4}{9}$$

One of the most fundamental concepts of fractions is their ability to be manipulated by multiplication or division. This is possible since $\frac{n}{n} = 1$ for any non-zero integer. As a result, multiplying or dividing by $\frac{n}{n}$ will not alter the original fraction since any number multiplied or divided by 1 doesn't change the value of that number. Fractions of the same value are known as equivalent fractions. For example, $\frac{2}{4}, \frac{4}{8}, \frac{50}{100}$, and $\frac{75}{150}$ are equivalent, as they all equal $\frac{1}{2}$.

To multiply two fractions, multiply the numerators to get the new numerator as well as multiply the denominators to get the new denominator. For example, $\frac{3}{5} \times \frac{2}{7} = \frac{3 \times 2}{5 \times 7} = \frac{6}{35}$.

Switching the numerator and denominator is called taking the *reciprocal* of a fraction. So the reciprocal of $\frac{4}{5}$ is $\frac{5}{4}$.

To divide one fraction by another, multiply the first fraction by the reciprocal of the second. So $\frac{3}{4} \div \frac{2}{5} = \frac{3}{4} \times \frac{5}{2} = \frac{15}{8}$.

If the numerator is smaller than the denominator, the fraction is a *proper fraction*. Otherwise, the fraction is said to be *improper*.

A *mixed number* is a number that is an integer plus some proper fraction, and is written with the integer first and the proper fraction to the right of it. Any mixed number can be written as an improper fraction by multiplying the integer by the denominator, adding the product to the value of the numerator, and dividing the sum by the original denominator. For example, $3\frac{1}{2} = \frac{3 \times 2 + 1}{2} = \frac{7}{2}$. Whole numbers can also be converted into fractions by placing the whole number as the numerator and making the denominator 1. For example, $3 = \frac{3}{1}$.

Percentages

Think of percentages as fractions with a denominator of 100. In fact, percentage means "per hundred." Problems often require converting numbers from percentages, fractions, and decimals. The following explains how to work through those conversions.

Converting Fractions to Percentages: Convert the fraction by using an equivalent fraction with a denominator of 100. For example, $\frac{3}{4} = \frac{3}{4} \times \frac{25}{25} = \frac{75}{100} = 75\%$

Converting Percentages to Fractions: Percentages can be converted to fractions by turning the percentage into a fraction with a denominator of 100. Be wary of questions asking the converted fraction to be written in the simplest form. For example, $35\% = \frac{35}{100}$ which, although correctly written, has a numerator and denominator with a greatest common factor of 5 and can be simplified to $\frac{7}{20}$.

Converting Percentages to Decimals: As a percentage is based on "per hundred," decimals and percentages can be converted by multiplying or dividing by 100. Practically speaking, this always amounts to moving the decimal point two places to the right or left, depending on the conversion. To convert a percentage to a decimal, move the decimal point two places to the left and remove the % sign. To convert a decimal to a percentage, move the decimal point two places to the right and add a "%" sign. Here are some examples:

65% = 0.65
0.33 = 33%
0.215 = 21.5%
99.99% = 0.9999
500% = 5.00
7.55 = 755%

Questions dealing with percentages can be difficult when they are phrased as word problems. These word problems almost always come in three varieties. The first type will ask to find what percentage of some number will equal another number. The second asks to determine what number is some percentage of another given number. The third will ask what number another number is a given percentage of.

One of the most important parts of correctly answering percentage word problems is to identify the numerator and the denominator. This fraction can then be converted into a percentage, as described above.

The following word problem shows how to make this conversion:

A department store carries several different types of footwear. The store is currently selling 8 athletic shoes, 7 dress shoes, and 5 sandals. What percentage of the store's footwear are sandals?

First, calculate what serves as the "whole," as this will be the denominator. How many total pieces of footwear does the store sell? The store sells 20 different types (8 athletic + 7 dress + 5 sandals).

Second, what footwear type is the question specifically asking about? Sandals. Thus, 5 is the numerator.

Third, the resultant fraction must be expressed as a percentage. The first two steps indicate that $\frac{5}{20}$ of the footwear pieces are sandals. This fraction must now be converted into a percentage:

$$\frac{5}{20} \times \frac{5}{5} = \frac{25}{100} = 25\%$$

Ratios and Proportions

Ratios are used to show the relationship between two quantities. The ratio of oranges to apples in the grocery store may be 3 to 2. That means that for every 3 oranges, there are 2 apples. This comparison can be expanded to represent the actual number of oranges and apples. Another example may be the number of boys to girls in a math class. If the ration of boys to girls is given as 2 to 5, that means there are 2 boys to every 5 girls in the class. Ratios can also be compared if the units in each ratio are the same. The ratio of boys to girls in the math class can be compared to the ratio of boys to girls in a science class by stating which ratio is higher and which is lower.

Rates are used to compare two quantities with different units. *Unit rates* are the simplest form of rate. With unit rates, the denominator in the comparison of two units is one. For example, if someone can type at a rate of 1000 words in 5 minutes, then his or her unit rate for typing is $\frac{1000}{5} = 200$ words in one minute or 200 words per minute. Any rate can be converted into a unit rate by dividing to make the denominator one. 1000 words in 5 minutes has been converted into the unit rate of 200 words per minute.

Ratios and rates can be used together to convert rates into different units. For example, if someone is driving 50 kilometers per hour, that rate can be converted into miles per hour by using a ratio known as the *conversion factor*. Since the given value contains kilometers and the final answer needs to be in miles, the ratio relating miles to kilometers needs to be used. There are 0.62 miles in 1 kilometer. This, written as a ratio and in fraction form, is

$$\frac{0.62\ miles}{1\ km}$$

To convert 50km/hour into miles per hour, the following conversion needs to be set up:

$$\frac{50\ km}{hour} * \frac{0.62\ miles}{1\ km} = 31\ miles\ per\ hour$$

The ratio between two similar geometric figures is called the *scale factor*. For example, a problem may depict two similar triangles, A and B. The scale factor from the smaller triangle A to the larger triangle B is given as 2 because the length of the corresponding side of the larger triangle, 16, is twice the corresponding side on the smaller triangle, 8. This scale factor can also be used to find the value of a missing side, x, in triangle A. Since the scale factor from the smaller triangle (A) to larger one (B) is 2, the larger corresponding side in triangle B (given as 25), can be divided by 2 to find the missing side in A (x = 12.5). The scale factor can also be represented in the equation $2A = B$ because two times the lengths of A gives the corresponding lengths of B. This is the idea behind similar triangles.

Much like a scale factor can be written using an equation like $2A = B$, a *relationship* is represented by the equation $Y = kX$. X and Y are proportional because as values of X increase, the values of Y also increase. A relationship that is inversely proportional can be represented by the equation $Y = \frac{k}{X}$, where the value of Y decreases as the value of x increases and vice versa.

Proportional reasoning can be used to solve problems involving ratios, percentages, and averages. Ratios can be used in setting up proportions and solving them to find unknowns. For example, if student completes an average of 10 pages of math homework in 3 nights, how long would it take the student to complete 22 pages? Both ratios can be written as fractions. The second ratio would contain the unknown.

The following proportion represents this problem, where x is the unknown number of nights:

$$\frac{10\ pages}{3\ nights} = \frac{22\ pages}{x\ nights}$$

Solving this proportion entails cross-multiplying and results in the following equation: $10x = 22 * 3$. Simplifying and solving for x results in the exact solution: $x = 6.6\ nights$. The result would be rounded up to 7 because the homework would be actually be completed on the 7th night.

The following problem uses ratios involving percentages:

If 20% of the class is girls and 30 students are in the class, how many girls are in the class?

To set up this problem, it is helpful to use the common proportion:

$$\frac{\%}{100} = \frac{is}{of}$$

Within the proportion, % is the percentage of girls, 100 is the total percentage of the class, *is* is the number of girls, and *of* is the total number of students in the class. Most percentage problems can be written using this language. To solve this problem, the proportion should be set up as $\frac{20}{100} = \frac{x}{30}$, and then solved for x. Cross-multiplying results in the equation $20 * 30 = 100x$, which results in the solution $x = 6$. There are 6 girls in the class.

Problems involving volume, length, and other units can also be solved using ratios. A problem may ask for the volume of a cone to be found that has a radius, $r = 7m$ and a height, $h = 16m$. Referring to the formulas provided on the test, the volume of a cone is given as:

$$V = \pi r^2 \frac{h}{3}$$

r is the radius, and h is the height. Plugging $r = 7$ and $h = 16$ into the formula, the following is obtained:

$$V = \pi(7^2)\frac{16}{3}$$

Therefore, volume of the cone is found to be approximately 821m^3. Sometimes, answers in different units are sought. If this problem wanted the answer in liters, 821m^3 would need to be converted.

Using the equivalence statement 1m^3 = 1000L, the following ratio would be used to solve for liters:

$$821\text{m}^3 * \frac{1000L}{1m^3}$$

Cubic meters in the numerator and denominator cancel each other out, and the answer is converted to 821,000 liters, or $8.21 * 10^5$ L.

Other conversions can also be made between different given and final units. If the temperature in a pool is 30°C, what is the temperature of the pool in degrees Fahrenheit? To convert these units, an equation is used relating Celsius to Fahrenheit. The following equation is used:

$$T_{°F} = 1.8T_{°C} + 32$$

Plugging in the given temperature and solving the equation for T yields the result:

$$T_{°F} = 1.8(30) + 32 = 86°F$$

Both units in the metric system and U.S. customary system are widely used.

Basic Geometry Relationships

The basic unit of geometry is a point. A point represents an exact location on a plane, or flat surface. The position of a point is indicated with a dot and usually named with a single uppercase letter, such as point A or point T. A point is a place, not a thing, and therefore has no dimensions or size. A set of points that lies on the same line is called collinear. A set of points that lies on the same plane is called coplanar.

The image above displays point *A*, point *B*, and point *C*.

A line is as series of points that extends in both directions without ending. It consists of an infinite number of points and is drawn with arrows on both ends to indicate it extends infinitely. Lines can be named by two points on the line or with a single, cursive, lower case letter. The two lines below could be named line *AB* or line *BA* or $\overleftrightarrow{AB}$ or $\overleftrightarrow{BA}$; and line *m*.

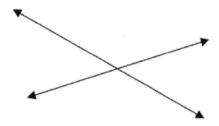

Two lines are considered parallel to each other if, while extending infinitely, they will never intersect (or meet). Parallel lines point in the same direction and are always the same distance apart. Two lines are considered perpendicular if they intersect to form right angles. Right angles are 90°. Typically, a small box is drawn at the intersection point to indicate the right angle.

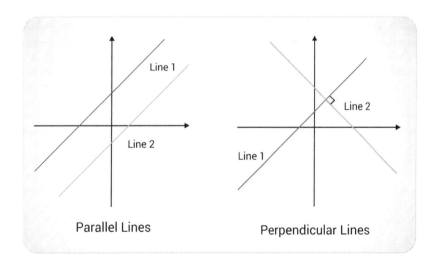

Line 1 is parallel to line 2 in the left image and is written as line 1 || line 2. Line 1 is perpendicular to line 2 in the right image and is written as line 1 ⊥ line 2.

A ray has a specific starting point and extends in one direction without ending. The endpoint of a ray is its starting point. Rays are named using the endpoint first, and any other point on the ray. The following ray can be named ray *AB* and written $\overrightarrow{AB}$.

A line segment has specific starting and ending points. A line segment consists of two endpoints and all the points in between. Line segments are named by the two endpoints. The example below is named segment *KL* or segment *LK*, written $\overline{KL}$ or $\overline{LK}$.

Classification of Angles

An angle consists of two rays that have a common endpoint. This common endpoint is called the vertex of the angle. The two rays can be called sides of the angle. The angle below has a vertex at point *B* and the sides consist of ray *BA* and ray *BC*. An angle can be named in three ways:

1. Using the vertex and a point from each side, with the vertex letter in the middle.
2. Using only the vertex. This can only be used if it is the only angle with that vertex.
3. Using a number that is written inside the angle.

The angle below can be written ∠*ABC* (read angle *ABC*), ∠*CBA*, ∠*B*, or ∠1.

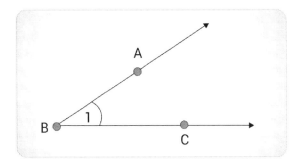

An angle divides a plane, or flat surface, into three parts: the angle itself, the interior (inside) of the angle, and the exterior (outside) of the angle. The figure below shows point *M* on the interior of the angle and point *N* on the exterior of the angle.

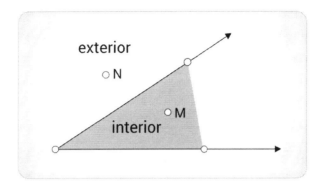

Angles can be measured in units called degrees, with the symbol °. The degree measure of an angle is between 0° and 180° and can be obtained by using a protractor.

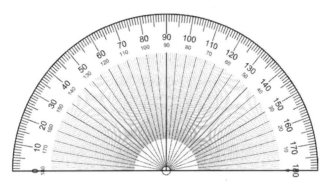

A straight angle (or simply a line) measures exactly 180°. A right angle's sides meet at the vertex to create a square corner. A right angle measures exactly 90° and is typically indicated by a box drawn in the interior of the angle. An acute angle has an interior that is narrower than a right angle. The measure of an acute angle is any value less than 90° and greater than 0°. For example, 89.9°, 47°, 12°, and 1°. An obtuse angle has an interior that is wider than a right angle. The measure of an obtuse angle is any value greater than 90° but less than 180°. For example, 90.1°, 110°, 150°, and 179.9°.

- Acute angles: Less than 90°
- Obtuse angles: Greater than 90°
- Right angles: 90°
- Straight angles: 180°

If two angles add together to give 90°, the angles are *complementary*.

If two angles add together to give 180°, the angles are *supplementary*.

When two lines intersect, the pairs of angles they form are always supplementary. The two angles marked here are supplementary:

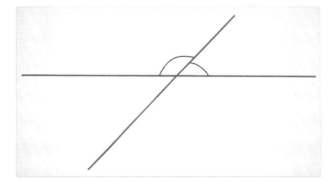

When two supplementary angles are next to one another or "adjacent" in this way, they always give rise to a straight line.

A polygon is a closed geometric figure in a plane (flat surface) consisting of at least 3 sides formed by line segments. These are often defined as two-dimensional shapes. Common two-dimensional shapes include circles, triangles, squares, rectangles, pentagons, and hexagons. Note that a circle is a two-dimensional shape without sides.

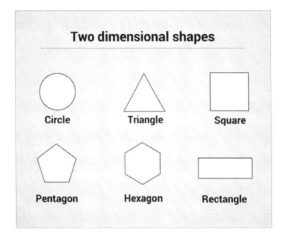

Polygons can be classified by the number of sides (also equal to the number of angles) they have. The following are the names of polygons with a given number of sides or angles:

# of sides	3	4	5	6	7	8	9	10
Name of polygon	Triangle	Quadrilateral	Pentagon	Hexagon	Septagon (or heptagon)	Octagon	Nonagon	Decagon

Triangles can be further classified by their sides and angles. A triangle with its largest angle measuring 90° is a right triangle. A triangle with the largest angle less than 90° is an acute triangle. A triangle with the largest angle greater than 90° is an obtuse triangle. Below is an example of a right triangle.

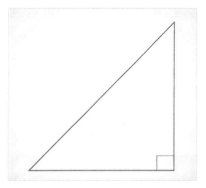

A triangle consisting of two equal sides and two equal angles is an isosceles triangle. A triangle with three equal sides and three equal angles is an equilateral triangle. A triangle with no equal sides or angles is a scalene triangle.

The three angles inside the triangle are called *interior angles* and add up to 180°.

For any triangle, the *Triangle Inequality Theorem* says that the following holds true: $A + B > C, A + C > B, B + C > A$. In addition, the sum of two angles must be less than 180°.

If two triangles have angles that agree with one another, that is, the angles of the first triangle are equal to the angles of the second triangle, then the triangles are called *similar*. Similar triangles look the same, but one can be a "magnification" of the other.

Two triangles with sides that are the same length must also be similar triangles. In this case, such triangles are called *congruent*. Congruent triangles have the same angles and lengths, even if they are rotated from one another.

Quadrilaterals can be further classified according to their sides and angles. A quadrilateral with exactly one pair of parallel sides is called a trapezoid. A quadrilateral that shows both pairs of opposite sides parallel is a parallelogram. Parallelograms include rhombuses, rectangles, and squares. A rhombus has four equal sides. A rectangle has four equal angles (90° each). A square has four 90° angles and four equal sides. Therefore, a square is both a rhombus and a rectangle.

A solid figure, or simply solid, is a figure that encloses a part of space. Some solids consist of flat surfaces only while others include curved surfaces. Solid figures are often defined as three-dimensional shapes. Common three-dimensional shapes include spheres, prisms, cubes, pyramids, cylinders, and cones.

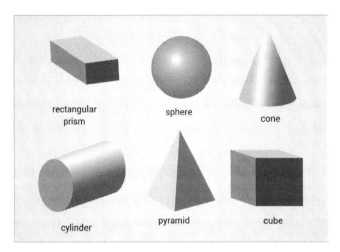

Perimeter is the measurement of a distance around something. It can be thought of as the length of the boundary, like a fence. It is found by adding together the lengths of all of the sides of a figure. Since a square has four equal sides, its perimeter can be calculated by multiplying the length of one side by 4. Thus, the formula is $P = 4 \times s$, where s equals one side. Like a square, a rectangle's perimeter is measured by adding together all of the sides. But as the sides are unequal, the formula is different. A rectangle has equal values for its lengths (long sides) and equal values for its widths (short sides), so the perimeter formula for a rectangle is $P = l + l + w + w = 2l + 2w$, where l is length and w is width. Perimeter is measured in simple units such as inches, feet, yards, centimeters, meters, miles, etc.

In contrast to perimeter, area is the space occupied by a defined enclosure, like a field enclosed by a fence. It is measured in square units such as square feet or square miles. Here are some formulas for the areas of basic planar shapes:

1. The area of a rectangle is $l \times w$, where w is the width and l is the length
2. The area of a square is s^2, where s is the length of one side (this follows from the formula for rectangles)
3. The area of a triangle with base b and height h is $\frac{1}{2}bh$
4. The area of a circle with radius r is πr^2

Volume is the measurement of how much space an object occupies, like how much space is in the cube. Volume questions will typically ask how much of something is needed to completely fill the object. It is measured in cubic units, such as cubic inches. Here are some formulas for the volumes of basic three-dimensional geometric figures:

1. For a regular prism whose sides are all rectangles, the volume is $l \times w \times h$, where w is the width, l is the length, and h is the height of the prism
2. For a cube, which is a prism whose faces are all squares of the same size, the volume is s^3
3. The volume of a sphere of radius r is given by $\frac{4}{3}\pi r^3$
4. The volume of a cylinder whose base has a radius of r and which has a height of h is given by $\pi r^2 h$

Word Problems

Word problems can appear daunting, but don't let the verbiage psyche you out. No matter the scenario or specifics, the key to answering them is to translate the words into a math problem. Always keep in mind what the question is asking and what operations could lead to that answer.

Translating Words into Math

When asked to rewrite a mathematical expression as a situation or translate a word problem into an expression, look for a series of key words indicating addition, subtraction, multiplication, or division:

Addition: add, altogether, together, plus, increased by, more than, in all, sum, and total

Subtraction: minus, less than, difference, decreased by, fewer than, remain, and take away

Multiplication: *times, twice, of, double,* and *triple*

Division: divided by, cut up, half, quotient of, split, and shared equally

Identifying and utilizing the proper units for the scenario requires knowing how to apply the conversion rates for money, length, volume, and mass. For example, given a scenario that requires subtracting 8 inches from $2\frac{1}{2}$ feet, both values should first be expressed in the same unit (they could be expressed $\frac{2}{3}$ft & $2\frac{1}{2}$ft, or 8in and 30in). The desired unit for the answer may also require converting back to another unit.

Consider the following scenario: A parking area along the river is only wide enough to fit one row of cars and is $\frac{1}{2}$ kilometers long. The average space needed per car is 5 meters. How many cars can be parked along the river? First, all measurements should be converted to similar units: $\frac{1}{2}$km = 500m. The operation(s) needed should be identified. Because the problem asks for the number of cars, the total space should be divided by the space per car. 500 meters divided by 5 meters per car yields a total of 100 cars. Written as an expression, the meters unit cancels and the cars unit is left: $\frac{500m}{5m/car}$ the same as $500m \times \frac{1\ car}{5m}$ yields 100 cars.

When dealing with problems involving elapsed time, breaking the problem down into workable parts is helpful. For example, suppose the length of time between 1:15pm and 3:45pm must be determined. From 1:15pm to 2:00pm is 45 minutes (knowing there are 60 minutes in an hour). From 2:00pm to 3:00pm is 1 hour. From 3:00pm to 3:45pm is 45 minutes. The total elapsed time is 45 minutes plus 1 hour plus 45 minutes. This sum produces 1 hour and 90 minutes. 90 minutes is over an hour, so this is converted to 1 hour (60 minutes) and 30 minutes. The total elapsed time can now be expressed as 2 hours and 30 minutes.

Example 1
Alexandra made $96 during the first 3 hours of her shift as a temporary worker at a law office. She will continue to earn money at this rate until she finishes in 5 more hours. How much does Alexandra make per hour? How much will Alexandra have made at the end of the day?

The hourly rate can be figured by dividing $96 by 3 hours to get $32 per hour. Now her total pay can be figured by multiplying $32 per hour by 8 hours, which comes out to $256.

Example 2
Bernard wishes to paint a wall that measures 20 feet wide by 8 feet high. It costs $0.10 to paint 1 square foot. How much money will Bernard need for paint?

The final quantity to compute is the *cost* to paint the wall. This will be ten cents ($0.10) for each square foot of area needed to paint. The area to be painted is unknown, but the dimensions of the wall are given; thus, it can be calculated.

The dimensions of the wall are 20 feet wide and 8 feet high. Since the area of a rectangle is length multiplied by width, the area of the wall is $8 \times 20 = 160$ square feet. Multiplying 0.1 x 160 yields $16 as the cost of the paint.

Data Analysis

Representing Data
Most statistics involve collecting a large amount of data, analyzing it, and then making decisions based on previously known information. These decisions also can be measured through additional data collection and then analyzed. Therefore, the cycle can repeat itself over and over. Representing the data visually is a large part of the process, and many plots on the real number line exist that allow this to be done. For example, a *dot plot* uses dots to represent data points above the number line. Also, a *histogram* represents a data set as a collection of rectangles, which illustrate the frequency distribution of the data. Finally, a *box plot* (also known as a *box and whisker plot*) plots a data set on the number line by segmenting the distribution into four quartiles that are divided equally in half by the median.

Here's an example of a box plot, a histogram, and a dot plot for the same data set:

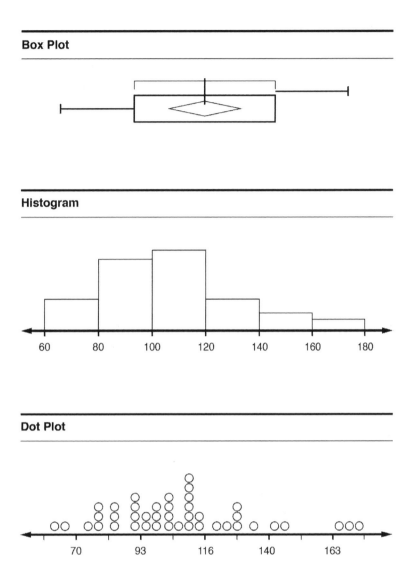

Box Plot

Histogram

Dot Plot

Comparing Data

Comparing data sets within statistics can mean many things. The first way to compare data sets is by looking at the center and spread of each set. The center of a data set can mean two things: median or mean. The *median* is the value that's halfway into each data set, and it splits the data into two intervals. The *mean* is the average value of the data within a set. It's calculated by adding up all of the data in the set and dividing the total by the number of data points. Outliers can significantly impact the mean. Additionally, two completely different data sets can have the same mean. For example, a data set with values ranging from 0 to 100 and a data set with values ranging from 44 to 56 can both have means of 50. The first data set has a much wider range, which is known as the *spread* of the data. This measures how varied the data is within each set. Spread can be defined further as either interquartile range or standard deviation. The *interquartile range (IQR)* is the range of the middle 50 percent of the data set. This range can be seen in the large rectangle on a box plot. The *standard*

45

deviation (σ) quantifies the amount of variation with respect to the mean. A lower standard deviation shows that the data set doesn't differ greatly from the mean. A larger standard deviation shows that the data set is spread out farther from the mean. The formula for standard deviation is:

$$\sigma = \sqrt{\frac{\sum(x - \bar{x})^2}{n - 1}}$$

x is each value in the data set, $\bar{x}$ is the mean, and n is the total number of data points in the set.

<u>Interpreting Data</u>
The shape of a data set is another way to compare two or more sets of data. If a data set isn't symmetric around its mean, it's said to be *skewed*. If the tail to the left of the mean is longer, it's said to be *skewed to the left*. In this case, the mean is less than the median. Conversely, if the tail to the right of the mean is longer, it's said to be *skewed to the right* and the mean is greater than the median. When classifying a data set according to its shape, its overall *skewness* is being discussed. If the mean and median are equal, the data set isn't skewed; it is *symmetric*.

An *outlier* is a data point that lies a great distance away from the majority of the data set. It also can be labelled as an *extreme value*. Technically, an outlier is any value that falls 1.5 times the IQR above the upper quartile or 1.5 times the IQR below the lower quartile. The effect of outliers in the data set is seen visually because they affect the mean. If there's a large difference between the mean and mode, outliers are the cause. The mean shows bias towards the outlying values. However, the median won't be affected as greatly by outliers.

<u>Normal Distribution</u>
A *normal distribution* of data follows the shape of a bell curve and the data set's median, mean, and mode are equal. Therefore, 50 percent of its values are less than the mean and 50 percent are greater than the mean. Data sets that follow this shape can be generalized using normal distributions. Normal distributions are described as *frequency distributions* in which the data set is plotted as percentages rather than true data points. A *relative frequency distribution* is one where the y-axis is between 0 and 1, which is the same as 0% to 100%. Within a standard deviation, 68 percent of the values are within 1 standard deviation of the mean, 95 percent of the values are within 2 standard deviations of the mean, and 99.7 percent of the values are within 3 standard deviations of the mean. The number of standard deviations that a data point falls from the mean is called the *z-score*. The formula for the z-score is $Z = \frac{x - \mu}{\sigma}$, where μ is the mean, σ is the standard deviation, and x is the data point. This formula is used to fit any data set that resembles a normal distribution to a standard normal distribution, in a process known as *standardizing*.

Here is a normal distribution with labelled z-scores:

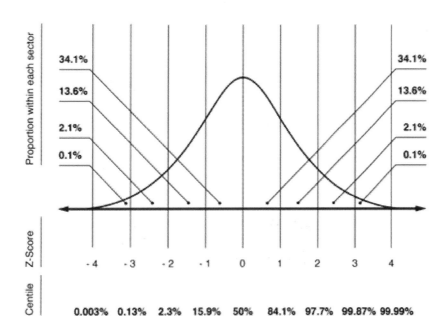

Population percentages can be estimated using normal distributions. For example, the probability that a data point will be less than the mean, or that the z-score will be less than 0, is 50%. Similarly, the probability that a data point will be within 1 standard deviation of the mean, or that the z-score will be between -1 and 1, is about 68.2%. When using a table, the left column states how many standard deviations (to one decimal place) away from the mean the point is, and the row heading states the second decimal place. The entries in the table corresponding to each column and row give the probability, which is equal to the area.

Areas Under the Curve

The area under the curve of a standard normal distribution is equal to 1. Areas under the curve can be estimated using the z-score and a table. The area is equal to the probability that a data point lies in that region in decimal form. For example, the area under the curve from $z = -1$ to $z = 1$ is 0.682.

Practice Questions

1. If a car can go 300 miles in 4 hours, how far can it go in an hour and a half?
   a. 100 miles
   b. 112.5 miles
   c. 135.5 miles
   d. 150 miles
   e. 165.5 miles

Maybe A

2. At the store, Jan buys $90 of apples and oranges. Apples cost $1 each and oranges cost $2 each. If Jan buys the same number of apples as oranges, how many oranges did she buy?
   a. 20
   b. 25
   c. 30
   d. 35
   e. 40

3. What is the volume of a box with rectangular sides 5 feet long, 6 feet wide, and 3 feet high?
   a. 60 cubic feet
   b. 75 cubic feet
   c. 90 cubic feet
   d. 100 cubic feet
   e. 115 cubic feet

4. A train traveling 50 miles per hour takes a trip of 3 hours. If a map has a scale of 1 inch per 10 miles, how many inches apart are the train's starting point and ending point on the map?
   a. 10
   b. 12
   c. 13
   d. 14
   e. 15

Maybe B

5. A traveler takes an hour to drive to a museum, spends 3 hours and 30 minutes there, and takes half an hour to drive home. What percentage of this time was spent driving?
   a. 15%
   b. 30%
   c. 40%
   d. 50%
   e. 60%

6. A truck is carrying three cylindrical barrels. Their bases have a diameter of 2 feet and they have a height of 3 feet. What is the total volume of the three barrels in cubic feet?
   a. 3 π
   b. 9 π
   c. 12 π
   d. 15 π
   e. 36 π

7. Greg buys a $10 lunch with 5% sales tax. He leaves a $2 tip after his bill. How much money does he spend?
   a. $12
   b. $12.50
   c. $13
   d. $13.25
   e. $16

8. Marty wishes to save $150 over a 4-day period. How much must Marty save each day on average?
   a. $33.50
   b. $35
   c. $37.50
   d. $40
   e. $45.75

9. Bernard can make $80 per day. If he needs to make $300 and only works full days, how many days will this take?
   a. 2
   b. 3
   c. 4
   d. 5
   e. 6

10. A couple buys a house for $150,000. They sell it for $165,000. What percentage did the house's value increase?
   a. 10%
   b. 13%
   c. 15%
   d. 17%
   e. 19%

11. A school has 15 teachers and 20 teaching assistants. They have 200 students. What is the ratio of faculty to students?
   a. 3:20
   b. 4:17
   c. 11:54
   d. 3:2
   e. 7:40

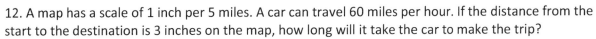

12. A map has a scale of 1 inch per 5 miles. A car can travel 60 miles per hour. If the distance from the start to the destination is 3 inches on the map, how long will it take the car to make the trip?
   a. 12 minutes
   b. 15 minutes
   c. 17 minutes
   d. 20 minutes
   e. 25 minutes

3 inches = 15 miles

13. Taylor works two jobs. The first pays $20,000 per year. The second pays $10,000 per year. She donates 15% of her income to charity. How much does she donate each year?

    a. $4500
    b. $5000
    c. $5500
    d. $6000
    e. $6500

30,000 × .15 = 4500

14. A box with rectangular sides is 24 inches wide, 18 inches deep, and 12 inches high. What is the volume of the box in cubic feet?

    a. 2
    b. 3
    c. 4
    d. 5
    e. 6

15. Kristen purchases $100 worth of CDs and DVDs. The CDs cost $10 each and the DVDs cost $15. If she bought four DVDs, how many CDs did she buy?

    a. One
    b. Two
    c. Three
    d. Four
    e. Five

15 × 4 = 60 DVD

16. Alan currently weighs 200 pounds, but he wants to lose weight to get down to 175 pounds. What is this difference in kilograms? (1 pound is approximately equal to 0.45 kilograms.)

    a. 9 kg
    b. 15.5 kg
    c. 78.75 kg
    d. 90 kg
    e. 11.25 kg

25 pounds × 0.45 kilograms

17. Johnny earns $2334.50 from his job each month. He pays $1437 for monthly expenses. Johnny is planning a vacation in 3 months' time that he estimates will cost $1750 total. How much will Johnny have left over from three months' of saving once he pays for his vacation?

    a. $948.50
    b. $584.50
    c. $852.50
    d. $942.50
    e. $984.50

2334.50 − 1437 = 897.50 × 3 = 2692.50 − 1750 = 942.50

18. The total perimeter of a rectangle is 36 cm. If the length of each side is 12 cm, what is the width?

    a. 3 cm
    b. 12 cm
    c. 6 cm
    d. 8 cm
    e. 9 cm

12 cm

36 − 24 = 12/2 = 6

12 cm

19. Dwayne has received the following scores on his math tests: 78, 92, 83, 97. What score must Dwayne get on his next math test to have an overall average of at least 90?
    a. 89
    b. 98
    c. 95
    d. 100
    e. 105

20. In Jim's school, there are 3 girls for every 2 boys. There are 650 students in total. Using this information, how many students are girls?
    a. 260
    b. 130
    c. 65
    d. 390
    e. 90

maybe E

21. Kimberley earns $10 an hour babysitting, and after 10 p.m., she earns $12 an hour, with the amount paid being rounded to the nearest hour accordingly. On her last job, she worked from 5:30 p.m. to 11 p.m. In total, how much did Kimberley earn on her last job?
    a. $45
    b. $57
    c. $62
    d. $42
    e. $67

6 hours   30 minutes

22. Keith's bakery had 252 customers go through its doors last week. This week, that number increased to 378. Express this increase as a percentage.
    a. 26%
    b. 50%
    c. 35%
    d. 12%
    e. 42%

23. For a group of 20 men, the median weight is 180 pounds and the range is 30 pounds. If each man gains 10 pounds, which of the following would be true?
    a. The median weight will increase, and the range will remain the same.
    b. The median weight and range will both remain the same.
    c. The median weight will stay the same, and the range will increase.
    d. The median weight and range will both increase.
    e. The median weight will decrease, and the range will remain the same.

24. Apples cost $2 each, while oranges cost $3 each. Maria purchased 10 fruits in total and spent $22. How many apples did she buy?
    a. 5
    b. 6
    c. 7
    d. 8
    e. 9

51

25. Five students take a test. The scores of the first four students are 80, 85, 75, and 60. If the median score is 80, which of the following could NOT be the score of the fifth student?

    a. 60
    b. 80
    c. 85
    d. 100
    e. 90

60 75   80   85

Answer Explanations

1. B: 300 miles in 4 hours is 300/4 = 75 miles per hour. In 1.5 hours, it will go 1.5×75 miles, or 112.5 miles

2. C: One apple/orange pair costs $3 total. Jan therefore bought 90/3 = 30 total pairs, and hence 30 oranges.

3. C: The volume of a box with rectangular sides is the length times width times height, so $5 \times 6 \times 3 = 90$ cubic feet.

4. E: First, the train's journey in the real word is $3 \times 50 = 150$ miles. On the map, 1 inch corresponds to 10 miles, so there is 150/10 = 15 inches on the map.

5. B: The total trip time is $1 + 3.5 + 0.5 = 5$ hours. The total time driving is $1 + 0.5 = 1.5$ hours. So the fraction of time spent driving is 1.5/5 or 3/10. To get the percentage, convert this to a fraction out of 100. The numerator and denominator are multiplied by 10, with a result of 30/100. The percentage is the numerator in a fraction out of 100, so 30%.

6. B: The volume of a cylinder is $\pi r^2 h$, where r is the radius and h is the height. The diameter is twice the radius, so these barrels have a radius of 1 foot. That means each barrel has a volume of $\pi 1^2 3 = 3\pi$ cubic feet. Since there are three of them, the total is $3 \times 3\pi = 9$ cubic feet.

7. B: The tip is not taxed, so he pays 5% tax only on the $10. 5% of $10 is $0.05 \times 10 = \$0.50$. Add up $10 + $2 + $0.50 to get $12.50.

8. C: The first step is to divide up $150 into four equal parts. 150/4 is 37.5, so she needs to save $37.50 per day on average.

9. C: 300/80 =30/8 = 15/4 =3.75. But Bernard is only working full days, so he will need to work 4 days, since 3 days is not sufficient.

10. A: The value went up by $165,000 − $150,000 = $15,000. Out of $150,000, this is $\frac{15,000}{150,000} = \frac{1}{10}$. Convert this to having a denominator of 100, the result is $\frac{10}{100}$ or 10%.

11. E: The total faculty is $15 + 20 = 35$. So the ratio is 35:200. Then, divide both of these numbers by 5, since 5 is a common factor to both, with a result of 7:40.

12. B: The journey will be $5 \times 3 = 15$ miles. A car travelling at 60 miles per hour is travelling at 1 mile per minute. So it will take 15/1 = 15 minutes to take the journey.

13. A: Taylor's total income is $20,000 + $10,000 = $30,000. 15% of this is $\frac{15}{100} = \frac{3}{20}$. So $\frac{3}{20} \times$ $30,000 = \frac{90,000}{20} = \frac{9000}{2} = \$4500$.

14. B: Since the answer will be in cubic feet rather than inches, start by converting from inches to feet for the dimensions of the box. There are 12 inches per foot, so the box is 24/12 = 2 feet wide, 18/12 = 1.5 feet deep, and 12/12 = 1 foot high. The volume is the product of these three together: $2 \times 1.5 \times 1 = 3$ cubic feet.

15. D: Kristen bought four DVDs, which would total a cost of $4 \times 15 = \$60$. She spent a total of $100, so she spent $100 – \$60 = \$40$ on CDs. Since they cost $10 each, she must have purchased $40/10 =$ four CDs.

16. E: Using the conversion rate, multiply the projected weight loss of 25 lb by $0.45 \frac{kg}{lb}$ to get the amount in kilograms (11.25 kg).

17. D: First, subtract $1437 from $2334.50 to find Johnny's monthly savings; this equals $897.50. Then, multiply this amount by 3 to find out how much he will have (in three months) before he pays for his vacation: this equals $2692.50. Finally, subtract the cost of the vacation ($1750) from this amount to find how much Johnny will have left: $942.50.

18. C: The formula for the perimeter of a rectangle is P=2L+2W, where P is the perimeter, L is the length, and W is the width. The first step is to substitute all of the data into the formula:

$$36 = 2(12) + 2W$$

Simplify by multiplying 2×12:

$$36 = 24 + 2W$$

Simplifying this further by subtracting 24 on each side, which gives:

$$36 - 24 = 24 - 24 + 2W$$
$$12 = 2W$$

Divide by 2:

$$6 = W$$

The width is 6 cm. Remember to test this answer by substituting this value into the original formula: $36 = 2(12) + 2(6)$.

19. D: To find the average of a set of values, add the values together and then divide by the total number of values. In this case, include the unknown value of what Dwayne needs to score on his next test, in order to solve it.

$$\frac{78 + 92 + 83 + 97 + x}{5} = 90$$

Add the unknown value to the new average total, which is 5. Then multiply each side by 5 to simplify the equation, resulting in:

$$78 + 92 + 83 + 97 + x = 450$$
$$350 + x = 450$$
$$x = 100$$

Dwayne would need to get a perfect score of 100 in order to get an average of at least 90.

Test this answer by substituting back into the original formula.

$$\frac{78 + 92 + 83 + 97 + 100}{5} = 90$$

20. D: Three girls for every two boys can be expressed as a ratio: 3:2. This can be visualized as splitting the school into 5 groups: 3 girl groups and 2 boy groups. The number of students that are in each group can be found by dividing the total number of students by 5:

650 divided by 5 equals 1 part, or 130 students per group

To find the total number of girls, multiply the number of students per group (130) by how the number of girl groups in the school (3). This equals 390, answer *D*.

21. C: Kimberley worked 4.5 hours at the rate of $10/h and 1 hour at the rate of $12/h. The problem states that her pay is rounded to the nearest hour, so the 4.5 hours would round up to 5 hours at the rate of $10/h. $(5h)(\$10/h) + (1h)(\$12/h) = \$50 + \$12 = \$62$.

22. B: First, calculate the difference between the larger value and the smaller value.

$$378 - 252 = 126$$

To calculate this difference as a percentage of the original value, and thus calculate the percentage *increase*, divide 126 by 252, then multiply by 100 to reach the percentage = 50%, answer *B*.

23. A: If each man gains 10 pounds, every original data point will increase by 10 pounds. Therefore, the man with the original median will still have the median value, but that value will increase by 10. The smallest value and largest value will also increase by 10 and, therefore, the difference between the two won't change. The range does not change in value and, thus, remains the same.

24. D: Let *a* be the number of apples and *o* the number of oranges. Then, the total cost is $2a + 3o = 22$, while it also known that $a + o = 10$. Using the knowledge of systems of equations, cancel the *o* variables by multiplying the second equation by -3. This makes the equation $-3a - 3o = -30$. Adding this to the first equation, the *o* values cancel to get $-a = -8$, which simplifies to *a* = 8.

25. A: Lining up the given scores provides the following list: 60, 75, 80, 85, and one unknown. Because the median needs to be 80, it means 80 must be the middle data point out of these five. Therefore, the unknown data point must be the fourth or fifth data point, meaning it must be greater than or equal to 80. The only answer that fails to meet this condition is 60.

Word Knowledge

Word Knowledge

Word knowledge is exactly what it sounds like: this portion of the exam is specifically constructed to test vocabulary skills and the ability to discern the best answer that matches the provided word. Unlike verbal analogies, which will test communication skills and problem-solving abilities along with vocabulary, word knowledge questions chiefly test vocabulary knowledge. While logic and reasoning come into play in this section, they are not as heavily emphasized as with the analogies. A prior knowledge of what the words mean is helpful in order to answer correctly. If the meaning of the words is unknown, that's fine, too; strategies should be used to rule out false answers and choose the correct ones. Here are some study strategies for an optimum performance.

Question Format

In contrast to the verbal analogies, word knowledge questions are very simple in construction. Instead of a comparison of words with an underlying connection, the prompt is just a single word. There are no special directions, alternate meanings, or analogies to work with. The objective is to analyze the given word and then choose the answer that means the same thing <u>or is closest in meaning</u> to the given word. Note the example below:

Blustery
    a. Hard
    b. Windy
    c. Mythical
    d. Stony
    e. Corresponding

All of the questions on the AFOQT word knowledge portion will appear exactly like the above sample. This is generally the standard layout throughout other exams, so some test-takers may already be familiar with the structure. The principle remains the same: at the top of the section, clear directions will be given to choose the answer that most precisely defines the given word. In this case, the answer is windy (B), since windy and blustery are synonymous.

Preparation

In truth, there is no set way to prepare for this portion of the exam that will guarantee a perfect score. This is simply because the words used on the test are unpredictable. There is no set list provided to study from. The definition of the provided word needs to be determined on the spot. This sounds challenging, but there are still ways to prepare mentally for this portion of the test. It may help to expand your vocabulary a little each day. Several resources are available, in books and online, that collect words and definitions that tend to show up frequently on standardized tests. Knowledge of words can increase the strength of your vocabulary.

Mindset is key. The meanings of challenging words can often be found by relying on the past experiences of the test-taker to help deduce the correct answer. How? Well, test-takers have been talking their entire lives—knowing words and how words work. It helps to have a positive mindset from the start. It's unlikely that all definitions of words will be known immediately, but the answer can still be found. There are aspects of words that are recognizable to help discern the correct answers and eliminate the incorrect ones. Below are some of the factors that contribute to word meanings.

Word Origins and Roots

Studying a foreign language in school, particularly Latin or any of the romance languages (Latin-influenced), is advantageous. English is a language highly influenced by Latin and Greek words. The roots of much of the English vocabulary have Latin origins; these roots can bind many words together and often allude to a shared definition. Here's an example:

Fervent
    a. Lame
    b. Joyful
    c. Thorough
    d. Boiling
    e. Cunning

Fervent descends from the Latin word, *fervere*, which means "to boil or glow" and figuratively means "impassioned." The Latin root present in the word is *ferv*, which is what gives fervent the definition: showing great warmth and spirit or spirited, hot, glowing. This provides a link to boiling (D) just by root word association, but there's more to analyze. Among the other choices, none relate to fervent. The word lame (A) means crippled, disabled, weak, or inadequate. None of these match with fervent. While being fervent can reflect joy, joyful (B) directly describes "a great state of happiness," while fervent is simply expressing the idea of having very strong feelings—not necessarily joy. Thorough (C) means complete, perfect, painstaking, or with mastery; while something can be done thoroughly and fervently, none of these words match fervent as closely as boiling does. Cunning (E) means crafty, deceiving or with ingenuity or dexterity. Doing something fervently does not necessarily mean it is done with dexterity. Not only does boiling connect in a linguistic way, but also in the way it is used in our language. While boiling can express being physically hot and undergoing a change, boiling is also used to reflect emotional states. People say they are "boiling over" when in heighted emotional states; "boiling mad" is another expression. Boiling, like fervent, also embodies a sense of heightened intensity. This makes boiling the best choice!

The Latin root *ferv* is seen in other words such as fervor, fervid, and even ferment. All of them are connected to and can be described by boil or glow, whether it is in a physical sense or in a metaphorical one. Such patterns can be seen in other word sets as well. Here's another example:

Gracious
    a. Fruitful
    b. Angry
    c. Grateful
    d. Understood
    e. Overheard

This one's a little easier; the answer is grateful (C), because both words mean thankful! Even if the meanings of both words are known, there's a connection found by looking at the beginnings of both words: *gra/grat*. Once again, these words are built on a root that stretches back to classical language. Both terms come from the Latin, *gratis*, which literally means "thanks."

Understanding root words can help identify the meaning in a lot of word choices, and help the test-taker grasp the nature of the given word. Many dictionaries, both in book form and online, offer information on the origins of words, which highlight these roots. When studying for the test, it helps to look up an unfamiliar word for its definition and then check to see if it has a root that can be connected to any other terms.

Pay Attention to Prefixes

The prefix of a word can actually reveal a lot about its definition. Many prefixes are actually Greco-Roman roots as well—but these are more familiar and a lot easier to recognize! When encountering any unfamiliar words, try looking at prefixes to discern the definition and then compare that with the choices. The prefix should be determined to help find the word's meaning. Here's an example question:

Premeditate
    a. Sporadic
    b. Calculated
    c. Interfere
    d. Determined
    e. Noble

With premeditate, there's the common prefix *pre*. This helps draw connections to other words like prepare or preassemble. *Pre* refers to "before, already being, or having already." Meditate means to think or plan. Premeditate means to think or plan beforehand with intent. Therefore, a term that deals with thinking or planning should be found, but also something done in preparation. Among the word choices, noble (E) and determined (D) are both adjectives with no hint of being related to something done before or in preparation. These choices are wrong. Sporadic (A) refers to events happening in irregular patterns, so this is quite the opposite of premeditated. Interfere (C) also has nothing to do with premeditate; it goes counter to premeditate in a way similar to sporadic. Calculated (B), however, fits! A route and the cost of starting a plan can be calculated. Calculated refers to acting with a full awareness of consequences, so inherently planning is involved. In fact, calculated is synonymous with premeditated, thus making it the correct choice. Just by paying attention to a prefix, the doors to a meaning can open to help easily figure out which word would be the best choice. Here's another example.

Regain
    a. Erupt
    b. Ponder
    c. Seek
    d. Recoup
    e. Enamor

Recoup (D) is the right answer. The prefix *re* often appears in front of words to give them the meaning of occurring again. Regain means to repossess something that was lost. Recoup, which also has the *re* prefix, literally means to regain. In this example, both the given word and the answer share the *re* prefix, which makes the pair easy to connect. However, don't rely *only* on prefixes to choose an answer. Make sure to analyze all of the options before marking an answer. Going through the other words in this sample, none of them come close to meaning regain except recoup. After checking to make sure that recoup is the best matching word, then mark it!

Positive Versus Negative Sounding Words

Another tool for the mental toolbox is simply distinguishing whether a word has a positive or negative connotation. Like electrical wires, words carry energy; they are crafted to draw certain attention and to have certain strength to them. Words can be described as positive and uplifting (a stronger word) or they can be negative and scathing (a stronger word). Sometimes they are neutral—having no particular connotation. Distinguishing how a word is supposed to be interpreted will not only help learn its definition, but also draw parallels with word choices. While it's true that words must usually be taken in the context of how they are used, word definitions have inherent meanings as well, meaning that they have a distinct vibe to pick up on. Here is an example.

Excellent
    a. Fair
    b. Optimum
    c. Reasonable
    d. Negative
    e. Agitation

As you know, excellent is a very positive word. It refers to something being better than good, or above average. In this sample, negative (D) and agitation (E) can easily be eliminated because these are both words with negative connotations. Reasonable (C) is more or less a neutral word: it's not bad but it doesn't communicate the higher quality that excellent represents. It's just, well, reasonable. This leaves the possible choices of fair (A) and optimum (B). Or does it? Fair *is* a positive word; it's used to describe things that are good, even beautiful. But in the modern context, fair is defined as good, but somewhat average or just decent: "You did a fairly good job" or, "That was fair." On the other hand, optimum is positive and a stronger word. Optimum describes the most favorable outcome. This makes optimum the best word choice that matches excellent in both strength and connotation. Not only are the two words positive, but they also express the same level of positivity! Here's another sample.

Repulse
    a. Draw
    b. Encumber
    c. Force
    d. Disgust
    e. Magnify

Repulse just sounds negative when said aloud. It is commonly used in the context of something being repulsive, disgusting, or that which is distasteful. It's also defined as an attack that drives people away. This tells us that we need a word that also carries a negative meaning. Magnify (E) is positive, while draw (A) and force (C) are both neutral. Encumber (B) and disgust (D) are negative. Disgust is a stronger negative than encumber. Of all the words given, only disgust directly defines a feeling of distaste and aversion that is synonymous with repulse and matches in both negativity and strength.

Parts of Speech

It is often very helpful to determine the part of speech of a word. Is it an adjective, adverb, noun, or verb, etc.? Often the correct answer will also be the same part of speech as the given word. Isolate the part of speech and what it describes and look for an answer choice that also describes the same part of speech. For example: if the given word is an adverb describing an action word, then look for another adverb describing an action word.

Swiftly
  a. Fast
  b. Quietly
  c. Angry
  d. Sudden
  e. Quickly

Swiftly is an adverb that describes the speed of an action. Angry (C), fast (A), and sudden (D) can be eliminated because they are not adverbs, and quietly (B) can be eliminated because it does not describe speed. This leaves quickly (E), which is the correct answer. Fast and sudden may throw off some test-takers because they both describe speed, but quickly matches more closely because it is an adverb, and swiftly is also an adverb.

Placing the Word in a Sentence

Often it is easier to discern the meaning of a word if it is used in a sentence. If the given word can be used in a sentence, then try replacing it with some of the answer choices to see which words seem to make sense in the same sentence. Here's an example.

Remarkable
  a. Often
  b. Capable
  c. Outstanding
  d. Shining
  e. Excluding

A sentence can be formed with the word remarkable. "My grade point average is remarkable." None of the examples make sense when replacing the word remarkable in the sentence other than the word outstanding (C), so outstanding is the obvious answer. Shining (D) is also a word with a positive connotation, but outstanding fits better in the sentence.

Looking for Relationships

Remember that all except one of the answer choices are wrong. If a close relationship between three or four of the answer choices can be found and not the fourth or fifth, then some of the choices can be eliminated. Sometimes all of the words are related except one; the one that is not related will often be the correct answer. Here is an example.

Outraged
  a. Angry
  b. Empty
  c. Forlorn
  d. Vacated
  e. Lonely

Notice that all of the answer choices have a negative connotation, but four of them are related to being alone or in low numbers. While two answer choices involve emotions—angry (A) and lonely (E), lonely is related to the other wrong answers, so angry is the best choice to match outraged.

Picking the Closest Answer

As the answer choices are reviewed, two scenarios might stand out. An exact definition match might not be found for the given word among the choices, or there are several word choices that can be considered synonymous to the given word. This is intentionally done to test the ability to draw parallels between the words in order to produce an answer that best fits the prompt word. Again, the closest fitting word will be the answer. Even when facing these two circumstances, finding the one word that fits best is the proper strategy. Here's an example:

Insubordination
- a. Cooperative
- b. Disciplined
- c. Rebel
- d. Contagious
- e. Wild

Insubordination refers to a defiance or utter refusal of authority. Looking over the choices, none of these terms provide definite matches to insubordination like insolence, mutiny, or misconduct would. This is fine; the answer doesn't have to be a perfect synonym. The choices don't reflect insubordination in any way, except rebel (C). After all, when rebel is used as a verb, it means to act against authority. It's also used as a noun: someone who goes against authority. Therefore, rebel is the best choice.

As with the verbal analogies section, playing the role of "detective" is the way to go as you may encounter two or even three answer choices that could be considered correct. However, the answer that best fits the prompt word's meaning is the best answer. Choices should be narrowed one word at a time. The least-connected word should be eliminated first and then proceed until one word is left that is the closest synonym.

Sequence
- a. List
- b. Range
- c. Series
- d. Replicate
- e. Iconic

A sequence reflects a particular order in which events or objects follow. The two closest options are list (A) and series (C). Both involve grouping things together, but which fits better? Consider each word more carefully. A list is comprised of items that fit in the same category, but that's really it. A list doesn't have to follow any particular order; it's just a list. On the other hand, a series is defined by events happening in a set order. A series relies on sequence, and a sequence can be described as a series. Thus, series is the correct answer!

Practice Questions

1. DEDUCE
    a. Explain
    b. Win
    c. Reason
    d. Gamble
    e. Undo

2. ELUCIDATE
    a. Learn
    b. Enlighten
    c. Plan
    d. Corroborate
    e. Conscious

3. VERIFY
    a. Criticize
    b. Change
    c. Teach
    d. Substantiate
    e. Resolve

4. INSPIRE
    a. Motivate
    b. Impale
    c. Exercise
    d. Patronize
    e. Collaborate

5. PERCEIVE
    a. Sustain
    b. Collect
    c. Prove
    d. Lead
    e. Comprehend

6. NOMAD
    a. Munching
    b. Propose
    c. Wanderer
    d. Conscientious
    e. Blissful

7. MALEVOLENT
   a. Evil
   b. Concerned
   c. Maximum
   d. Cautious
   e. Crazy

8. PERPLEXED
   a. Annoyed
   b. Vengeful
   c. Injured
   d. Confused
   e. Prepared

9. LYRICAL
   a. Whimsical
   b. Vague
   c. Fruitful
   d. Expressive
   e. Playful

10. BREVITY
   a. Dullness
   b. Dangerous
   c. Brief
   d. Ancient
   e. Calamity

11. IRATE
   a. Anger
   b. Knowledge
   c. Tired
   d. Confused
   e. Taciturn

12. LUXURIOUS
   a. Faded
   b. Bright
   c. Lavish
   d. Inconsiderate
   e. Overwhelming

13. IMMOBILE
   a. Fast
   b. Slow
   c. Eloquent
   d. Vivacious
   e. Sedentary

14. MENDACIOUS
    a. Earnest
    b. Bold
    c. Criminal
    d. Liar
    e. Humorous

15. CHIVALROUS
    a. Fierce
    b. Annoying
    c. Rude
    d. Dangerous
    e. Courteous

16. RETORT
    a. Conversation
    b. Jest
    c. Counter
    d. Flexible
    e. Erudite

17. SUBLIMINAL
    a. Subconscious
    b. Transportation
    c. Underground
    d. Substitute
    e. Penumbral

18. INCITE
    a. Understanding
    b. Illumination
    c. Rally
    d. Judgment
    e. Compose

19. MONIKER
    a. Name
    b. Mockery
    c. Umbrella
    d. Insult
    e. Burden

20. SERENDIPITOUS
    a. Creation
    b. Sympathy
    c. Unfortunate
    d. Calm
    e. Coincidental

21. OVERBEARING
    a. Neglect
    b. Overacting
    c. Clandestine
    d. Formidable
    e. Amicable

22. PREVENT
    a. Avert
    b. Rejoice
    c. Endow
    d. Fulfill
    e. Ensure

23. REPLENISH
    a. Falsify
    b. Hindsight
    c. Dwell
    d. Refresh
    e. Nominate

24. REGALE
    a. Remember
    b. Grow
    c. Outnumber
    d. Entertain
    e. Bore

25. ABATE
    a. Anger
    b. Forlorn
    c. Withdraw
    d. Excellent
    e. Crazed

Answer Explanations

1. C: To deduce something is to figure it out using reason. Although this might cause a win and prompt an explanation to further understanding, the art of deduction is logical reasoning.

2. B: To elucidate, a light is figuratively shined on a previously unknown or confusing subject. This Latin root, *lux* meaning *light*, prominently figures into the solution. Enlighten means to educate, or bring into the light.

3. D: Looking at the Latin word *veritas*, meaning *truth*, will yield a clue as to the meaning of verify. To verify is the act of finding or assessing the truthfulness of something. This usually means amassing evidence to substantiate a claim. Substantiate, of course, means to provide evidence to prove a point.

4. A: If someone is inspired, they are motivated to do something. Someone who is an inspiration motivates others to follow his or her example.

5. E: All the connotations of perceive involve the concept of seeing. Whether figuratively or literally, perceiving implies the act of understanding what is presented. Comprehending is synonymous with this understanding.

6. C: Nomadic tribes are those who, throughout history and even today, prefer to wander their lands instead of settling in any specific place. Wanderer best describes these people.

7. A: Malevolent literally means bad or evil-minded. The clue is also in the Latin root *mal-* that translates to bad.

8. D: Perplexed means baffled or puzzled, which are synonymous with confused.

9. D: Lyrical is used to refer to something being poetic or song-like, characterized by showing enormous imagination and description. While the context of lyrical can be playful or even whimsical, the best choice is expressive, since whatever emotion lyrical may be used to convey in context will be expressive in nature.

10. C: Brevity literally means brief or concise. Note the similar beginnings of brevity and brief—from the Latin *brevis*, meaning brief.

11. A: Irate means being in a state of anger. Clearly this is a negative word that can be paired with another word in kind. The closest word to match this is obviously anger. Research would also reveal that irate comes from the Latin *ira*, which means anger.

12. C: Lavish is a synonym for luxurious—both describe elaborate and/or elegant lifestyles and/or settings.

13. E: Immobile obviously means *not able to move*. The two best selections are *B* and *E*—but slow still implies some form of motion, whereas sedentary has the connotation of being seated and/or inactive for a significant portion of time and/or as a natural tendency.

14. D: Mendacious describes dishonesty or lying in several ways. This is another word of classical lineage. Mendacio in Latin means *liar*. While liar lacks the Latin root, the meanings fit.

15. E: Chivalrous reflects showing respect and courtesy toward others, particularly women.

16. C: Retort is a verb that means *to answer back*, usually in a sharp manner. This term embodies the idea of a response, emphasized by the *re-* prefix meaning *back, again*. While a retort is used in conversations and even as a jest, neither term embodies the idea of addressing someone again. Counter, however, means to respond in opposition when used as a verb.

17. A: Subliminal and subconscious share the Latin prefix *sub-*, meaning under or below, or more commonly used when talking about messages the sender doesn't want the receiver to consciously take note of. The word subliminal means beneath the consciousness. Thus, subconscious is the perfect match.

18. C: Although incite usually has negative connotations, leaders can incite their followers to benevolent actions as well. In both cases, people rally to support a cause.

19. A: While a moniker commonly refers to a title, this is technically a designated name. Monikers can be mockeries, insults, and/or burdens, but none of these are direct forms of a title or identification.

20. E: Events that occur through serendipity happen purely by chance. Serendipitous conveys the idea of something unplanned yet potentially desirable. Coincidental is defined as happening by chance.

21. B: Overbearing refers to domineering or being oppressive. This is emphasized in the *over* prefix, which emphasizes an excess in definitions. This prefix is also seen in overacting. Similar to overbearing, overacting reflects an excess of action.

22. A: The *pre* prefix describes something that occurs before an event. Prevent means to stop something before it happens. This leads to a word that relates to something occurring beforehand and, in a way, is preventive—wanting to stop something. Avert literally means to turn away or ward off an impending circumstance, making it the best fit.

23. D: Refresh is synonymous with replenish. Both words mean to restore or refill. Additionally, these terms do share the *re-* prefix as well.

24. D: Regale literally means to amuse someone with a story. This is a very positive word; the best way to eliminate choices is to look for a term that matches regale in both positive context/sound and definition. Entertain is both a positive word and a synonym of regale.

25. C: Abate is defined as something becoming less intense and fading. The only word that matches abate is withdraw, which means to go back or draw away from a particular position.

Math Knowledge

The Scope of the Math Knowledge Section

The Math Knowledge section of the test involves everything included in the Arithmetic Reasoning section, as well as some additional mathematical operations and techniques. It is, however, much less focused on word problems.

How to Prepare

Although this section of the test will be less focused on word problems, it is still very important to practice the types of problems in this section. As mentioned before, to really learn mathematics, it is important to practice and not just read through instructions. Approach this section the same as the Arithmetic Reasoning: first, read through the study guide here, then try the practice problems, and lastly, compare your answers with the solutions given below. You may utilize a slightly different method for solving a problem since there are sometimes multiple approaches that will work.

Exponents

An exponent is an operation used as shorthand for a number multiplied or divided by itself for a defined number of times.

$$3^7 = 3 \times 3 \times 3 \times 3 \times 3 \times 3 \times 3$$

In this example, the 3 is called the base, and the 7 is called the exponent. The exponent is typically expressed as a superscript number near the upper right side of the base, but can also be identified as the number following a caret symbol (^). This operation would be verbally expressed as "3 to the 7th power" or "3 raised to the power of 7." Common exponents are 2 and 3. A base raised to the power of 2 is referred to as having been "squared," while a base raised to the power of 3 is referred to as having been "cubed."

Several special rules apply to exponents. First, the Zero Power Rule finds that any number raised to the zero power equals 1. For example, 100^0, 2^0, $(-3)^0$ and 0^0 all equal 1 because the bases are raised to the zero power.

Second, exponents can be negative. With negative exponents, the equation is expressed as a fraction, as in the following example:

$$3^{-7} = \frac{1}{3^7} = \frac{1}{3 \times 3 \times 3 \times 3 \times 3 \times 3 \times 3}$$

Third, the Power Rule concerns exponents being raised by another exponent. When this occurs, the exponents are multiplied by each other:

$$(x^2)^3 = x^6 = (x^3)^2$$

Fourth, when multiplying two exponents with the same base, the Product Rule requires that the base remains the same and the exponents are added. For example, $a^x \times a^y = a^{x+y}$. Since addition and multiplication are commutative, the two terms being multiplied can be in any order.

$$x^3 x^5 = x^{3+5} = x^8 = x^{5+3} = x^5 x^3$$

Fifth, when dividing two exponents with the same base, the Quotient Rule requires that the base remains the same, but the exponents are subtracted. So, $a^x \div a^y = a^{x-y}$. Since subtraction and division are not commutative, the two terms must remain in order.

$$x^5 x^{-3} = x^{5-3} = x^2 = x^5 \div x^3 = \frac{x^5}{x^3}$$

Additionally, 1 raised to any power is still equal to 1, and any number raised to the power of 1 is equal to itself. In other words, $a^1 = a$ and $14^1 = 14$.

Exponents play an important role in scientific notation to present extremely large or small numbers as follows: $a \times 10^b$. To write the number in scientific notation, the decimal is moved until there is only one digit on the left side of the decimal point, indicating that the number a has a value between 1 and 10. The number of times the decimal moves indicates the exponent to which 10 is raised, here represented by b. If the decimal moves to the left, then b is positive, but if the decimal moves to the right, then b is negative. See the following examples:

$$3{,}050 = 3.05 \times 10^3$$

$$-777 = -7.77 \times 10^2$$

$$0.000123 = 1.23 \times 10^{-4}$$

$$-0.0525 = -5.25 \times 10^{-2}$$

Roots

The square root symbol is expressed as $\sqrt{\ }$ and is commonly known as the radical. Taking the root of a number is the inverse operation of multiplying that number by itself some amount of times. For example, squaring the number 7 is equal to 7×7, or 49. Finding the square root is the opposite of finding an exponent, as the operation seeks a number that when multiplied by itself equals the number in the square root symbol.

For example, $\sqrt{36}$ = 6 because 6 multiplied by 6 equals 36. Note, the square root of 36 is also -6 since $-6 \times -6 = 36$. This can be indicated using a plus/minus symbol like this: ±6. However, square roots are often just expressed as a positive number for simplicity with it being understood that the true value can be either positive or negative.

Perfect squares are numbers with whole number–square roots. The list of perfect squares begins with 0, 1, 4, 9, 16, 25, 36, 49, 64, 81, and 100.

Determining the square root of imperfect squares requires a calculator to reach an exact figure. It's possible, however, to approximate the answer by finding the two perfect squares that the number fits between. For example, the square root of 40 is between 6 and 7 since the squares of those numbers are 36 and 49, respectively.

Square roots are the most common root operation. If the radical doesn't have a number to the upper left of the symbol $\sqrt{\ }$, then it's a square root. Sometimes a radical includes a number in the upper left, like $\sqrt[3]{27}$, as in the other common root type—the cube root. Complicated roots like the cube root often require a calculator.

Parentheses

Parentheses separate different parts of an equation, and operations within them should be thought of as taking place before the outside operations take place. Practically, this means that the distinction between what is inside and outside of the parentheses decides the order of operations that the equation follows. Failing to solve operations inside the parentheses before addressing the part of the equation outside of the parentheses will lead to incorrect results.

For example, let's analyze $5 - (3 + 25)$. The addition operation within the parentheses must be solved first. So $3 + 25 = 28$, leaving $5 - (28) = -23$. If this was solved in the incorrect order of operations, the solution might be found to be $5 - 3 + 25 = 2 + 25 = 27$, which would be wrong.

Equations often feature multiple layers of parentheses. To differentiate them, square brackets [ ] and braces { } are used in addition to parentheses. The innermost parentheses must be solved before working outward to larger brackets. For example, in $\{2 \div [5 - (3 + 1)]\}$, solving the innermost parentheses $(3 + 1)$ leaves $\{2 \div [5 - (4)]\}$. $[5 - (4)]$ is now the next smallest, which leaves $\{2 \div [1]\}$ in the final step, and 2 as the answer.

Order of Operations

When solving equations with multiple operations, special rules apply. These rules are known as the Order of Operations. The order is as follows: Parentheses, Exponents, Multiplication and Division from left to right, and Addition and Subtraction from left to right. A popular pneumonic device to help remember the order is Please Excuse My Dear Aunt Sally (PEMDAS). Evaluate the following two problems to understand the Order of Operations:

1) $4 + (3 \times 2)^2 \div 4$

> First, solve the operation within the parentheses: $4 + 6^2 \div 4$.
> Second, solve the exponent: $4 + 36 \div 4$.
> Third, solve the division operation: $4 + 9$.
> Fourth, finish the operation with addition for the answer, 13.

2) $2 \times (6 + 3) \div (2 + 1)^2$

> $2 \times 9 \div (3)^2$
> $2 \times 9 \div 9$
> $18 \div 9$
> 2

Positive and Negative Numbers

Signs
Aside from 0, numbers can be either positive or negative. The sign for a positive number is the plus sign or the + symbol, while the sign for a negative number is minus sign or the − symbol. If a number has no designation, then it's assumed to be positive.

<u>Absolute Values</u>
Both positive and negative numbers are valued according to their distance from 0. Look at this number line for +3 and -3:

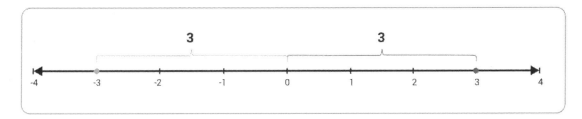

Both 3 and -3 are three spaces from 0. The distance from 0 is called its absolute value. Thus, both -3 and 3 have an absolute value of 3 since they're both three spaces away from 0.

An absolute number is written by placing | | around the number. So, |3| and |−3| both equal 3, as that's their common absolute value.

Implications for Addition and Subtraction

For addition, if all numbers are either positive or negative, simply add them together. For example, $4 + 4 = 8$ and $-4 + -4 = -8$. However, things get tricky when some of the numbers are negative and some are positive.

Take $6 + (-4)$ as an example. First, take the absolute values of the numbers, which are 6 and 4. Second, subtract the smaller value from the larger. The equation becomes $6 - 4 = 2$. Third, place the sign of the original larger number on the sum. Here, 6 is the larger number, and it's positive, so the sum is 2.

Here's an example where the negative number has a larger absolute value: $(-6) + 4$. The first two steps are the same as the example above. However, on the third step, the negative sign must be placed on the sum, as the absolute value of (-6) is greater than 4. Thus, $-6 + 4 = -2$.

The absolute value of numbers implies that subtraction can be thought of as flip the sign of the number following the subtraction sign and simply adding the two numbers. This means that subtracting a negative number will in fact be adding the positive absolute value of the negative number. Here are some examples:

$$-6 - 4 = -6 + -4 = -10$$

$$3 - -6 = 3 + 6 = 9$$

$$-3 - 2 = -3 + -2 = -5$$

Implications for Multiplication and Division

For multiplication and division, if both numbers are positive, then the product or quotient is always positive. If both numbers are negative, then the product or quotient is also positive. However, if the numbers have opposite signs, the product or quotient is always negative.

Simply put, the product in multiplication and quotient in division is always positive, unless the numbers have opposing signs, in which case it's negative.

Here are some examples:

$$(-6) \times (-5) = 30$$

$$(-50) \div 10 = -5$$

$$8 \times |-7| = 56$$

$$(-48) \div (-6) = 8$$

If there are more than two numbers in a multiplication problem, then whether the product is positive or negative depends on the number of negative numbers in the problem. If there is an odd number of negatives, then the product is negative. If there is an even number of negative numbers, then the result is positive.

Here are some examples:

$$(-6) \times 5 \times (-2) \times (-4) = -240$$

$$(-6) \times 5 \times 2 \times (-4) = 240$$

Polynomials

An expression of the form ax^n, where n is a non-negative integer, is called a *monomial* because it contains one term. A sum of monomials is called a *polynomial*. For example, $-4x^3 + x$ is a polynomial, while $5x^7$ is a monomial. A function equal to a polynomial is called a *polynomial function*.

The monomials in a polynomial are also called the *terms* of the polynomial.

The constants that precede the variables are called *coefficients*.

The highest value of the exponent of x in a polynomial is called the *degree* of the polynomial. So, $-4x^3 + x$ has a degree of 3, while $-2x^5 + x^3 + 4x + 1$ has a degree of 5. When multiplying polynomials, the degree of the result will be the sum of the degrees of the two polynomials being multiplied.

Addition and subtraction operations can be performed on polynomials with like terms. *Like terms* refer to terms that have the same variable and exponent. The two following polynomials can be added together by collecting like terms:

$$(x^2 + 3x - 4) + (4x^2 - 7x + 8)$$

The x^2 terms can be added as $x^2 + 4x^2 = 5x^2$. The x terms can be added as $3x + -7x = -4x$, and the constants can be added as $-4 + 8 = 4$. The following expression is the result of the addition:

$$5x^2 - 4x + 4$$

When subtracting polynomials, the same steps are followed, only subtracting like terms together.

To multiply two polynomials, each term of the first polynomial is multiplied by each term of the second polynomial and then the results are added.

For example: $(4x^2 + x)(-x^3 + x) = 4x^2(-x^3) + 4x^2(x) + x(-x^3) + x(x) = -4x^5 + 4x^3 - x^4 + x^2$.

In the case where each polynomial has two terms, like in this example, some students find it helpful to use the FOIL method. FOIL is a technique for generating polynomials through the multiplication of binomials. A polynomial is an expression of multiple variables (for example, x, y, z) in at least three terms involving only the four basic operations and exponents. FOIL is an acronym for First, Outer, Inner, and Last. *First* represents the multiplication of the terms appearing first in the binomials. *Outer* means multiplying the outermost terms. *Inner* means multiplying the terms inside. *Last* means multiplying the last terms of each binomial.

1) Simplify $(x + 10)(x + 4) = (x \times x) + (x \times 4) + (10 \times x) + (10 \times 4)$
$\quad\quad\quad\quad\quad\quad\quad\quad\quad\quad$ First $\quad$ Outer $\quad$ Inner $\quad$ Last

After multiplying these binomials, it's time to solve the operations and combine like terms. Thus, the expression becomes: $x^2 + 4x + 10x + 40 = x^2 + 14x + 40$

The process of *factoring* a polynomial means to write the polynomial as a product of other (generally simpler) polynomials. Here is an example: $x^2 - 4x + 3 = (x - 1)(x - 3)$. If a certain monomial divides every term of the polynomial, it gets factored out of each term.

For example:

$$4x^3 + 16x^2 = 4x^2(x + 4).$$

$$x^2 + 2xy + y^2 = (x + y)^2 \text{ or } x^2 - 2xy + y^2 = (x - y)^2$$

$$x^2 - y^2 = (x + y)(x - y)$$

$$x^3 + y^3 = (x + y)(x^2 - xy + y^2)$$

$$x^3 - y^3 = (x - y)(x^2 + xy + y^2)$$

$$x^3 + 3x^2y + 3xy^2 + y^3 = (x + y)^3 \text{ and } x^3 - 3x^2y + 3xy^2 - y^3 = (x - y)^3$$

Sometimes, it can be necessary to rewrite the polynomial in some clever way before applying the above rules.

Consider the problem of factoring $x^4 - 1$. This does not immediately look like any of the cases for which there are rules.

However, it's possible to think of this polynomial as: $x^4 - 1 = (x^2)^2 - (1^2)^2$, which can be simplified to: $(x^2)^2 - (1^2)^2 = (x^2 + 1^2)(x^2 - 1^2) = (x^2 + 1)(x^2 - 1)$.

Rational Expressions

A fraction, or ratio, wherein each part is a polynomial, defines *rational expressions*. Some examples include $\frac{2x+6}{x}$, $\frac{1}{x^2-4x+8}$, and $\frac{z^2}{x+5}$. Exponents on the variables are restricted to whole numbers, which means roots and negative exponents are not included in rational expressions.

Rational expressions can be transformed by factoring. For example, the expression $\frac{x^2-5x+6}{(x-3)}$ can be rewritten by factoring the numerator to obtain $\frac{(x-3)(x-2)}{(x-3)}$. Therefore, the common binomial $(x-3)$ can cancel so that the simplified expression is $\frac{(x-2)}{1} = (x-2)$.

Additionally, other rational expressions can be rewritten to take on different forms. Some may be factorable in themselves, while others can be transformed through arithmetic operations. Rational expressions are closed under addition, subtraction, multiplication, and division by a nonzero expression. *Closed* means that if any one of these operations is performed on a rational expression, the result will still be a rational expression. The set of all real numbers is another example of a set closed under all four operations.

Adding and subtracting rational expressions is based on the same concepts as adding and subtracting simple fractions. For both concepts, the denominators must be the same for the operation to take place. For example, here are two rational expressions:

$$\frac{x^3-4}{(x-3)} + \frac{x+8}{(x-3)}$$

Since the denominators are both $(x-3)$, the numerators can be combined by collecting like terms to form:

$$\frac{x^3+x+4}{(x-3)}$$

If the denominators are different, they need to be made common (the same) by using the Least Common Denominator (LCD). Each denominator needs to be factored, and the LCD contains each factor that appears in any one denominator the greatest number of times it appears in any denominator. The original expressions need to be multiplied times a form of 1, which will turn each denominator into the LCD. This process is like adding fractions with unlike denominators. It is also important when working with rational expressions to define what value of the variable makes the denominator zero. For this particular value, the expression is undefined.

Multiplication of rational expressions is performed like multiplication of fractions. The numerators are multiplied; then, the denominators are multiplied. The final fraction is then simplified. The expressions are simplified by factoring and cancelling out common terms. In the following example, the numerator of the second expression can be factored first to simplify the expression before multiplying:

$$\frac{x^2}{(x-4)} * \frac{x^2-x-12}{2}$$

$$\frac{x^2}{(x-4)} * \frac{(x-4)(x+3)}{2}$$

The $(x-4)$ on the top and bottom cancel out:

$$\frac{x^2}{1} * \frac{(x+3)}{2}$$

Then multiplication is performed, resulting i:

$$\frac{x^3 + 3x^2}{2}$$

Dividing rational expressions is similar to the division of fractions, where division turns into multiplying by a reciprocal. So the following expression can be rewritten as a multiplication problem:

$$\frac{x^2 - 3x + 7}{x - 4} \div \frac{x^2 - 5x + 3}{x - 4}$$

$$\frac{x^2 - 3x + 7}{x - 4} * \frac{x - 4}{x^2 - 5x + 3}$$

The $x - 4$ cancels out, leaving:

$$\frac{x^2 - 3x + 7}{x^2 - 5x + 3}$$

The final answers should always be completely simplified. If a function is composed of a rational expression, the zeros of the graph can be found from setting the polynomial in the numerator as equal to zero and solving. The values that make the denominator equal to zero will either exist on the graph as a hole or a vertical asymptote.

Linear Equations and Inequalities

Linear relationships describe the way two quantities change with respect to each other. The relationship is defined as linear because a line is produced if all the sets of corresponding values are graphed on a coordinate grid. When expressing the linear relationship as an equation, the equation is often written in the form $y = mx + b$ (slope-intercept form) where m and b are numerical values and x and y are variables (for example, $y = 5x + 10$). Given a linear equation and the value of either variable (x or y), the value of the other variable can be determined.

When graphing a linear equation, note that the ratio of the change of the y coordinate to the change in the x coordinate is constant between any two points on the resulting line, no matter which two points are chosen. In other words, in a pair of points on a line, (x_1, y_1) and (x_2, y_2), with $x_1 \neq x_2$ so that the two points are distinct, then the ratio $\frac{y_2 - y_1}{x_2 - x_1}$ will be the same, regardless of which particular pair of points are chosen. This ratio, $\frac{y_2 - y_1}{x_2 - x_1}$, is called the *slope* of the line and is frequently denoted with the letter m. If slope m is positive, then the line goes upward when moving to the right, while if slope m is negative, then the line goes downward when moving to the right. If the slope is 0, then the line is called *horizontal*, and the y coordinate is constant along the entire line. In lines where the x coordinate is constant along the entire line, y is not actually a function of x. For such lines, the slope is not defined. These lines are called *vertical* lines.

Linear functions may take forms other than $y = ax + b$. The most common forms of linear equations are explained below:

1. Standard Form: $Ax + By = C$, in which the slope is given by $m = \frac{-A}{B}$, and the y-intercept is given by $\frac{C}{B}$.

2. Slope-Intercept Form: $y = mx + b$, where the slope is m and the y intercept is b.

3. Point-Slope Form: $y - y_1 = m(x - x_1)$, where the slope is m and (x_1, y_1) is any point on the chosen line.

4. Two-Point Form: $\frac{y - y_1}{x - x_1} = \frac{y_2 - y_1}{x_2 - x_1}$, where (x_1, y_1) and (x_2, y_2) are any two distinct points on the chosen line. Note that the slope is given by $m = \frac{y_2 - y_1}{x_2 - x_1}$.

5. Intercept Form: $\frac{x}{x_1} + \frac{y}{y_1} = 1$, in which x_1 is the x-intercept and y_1 is the y-intercept.

These five ways to write linear equations are all useful in different circumstances. Depending on the given information, it may be easier to write one of the forms over another.

If $y = mx$, y is directly proportional to x. In this case, changing x by a factor changes y by that same factor. If $y = \frac{m}{x}$, y is inversely proportional to x. For example, if x is increased by a factor of 3, then y will be decreased by the same factor, 3.

The first steps to solving linear equations are distributing, if necessary, and combining any like terms on the same side of the equation. Sides of an equation are separated by an *equal* sign. Next, the equation is manipulated to show the variable on one side. Whatever is done to one side of the equation must be done to the other side of the equation to remain equal. Inverse operations are then used to isolate the variable and undo the order of operations backwards. Addition and subtraction are undone, then multiplication and division are undone.

For example, solve $4(t - 2) + 2t - 4 = 2(9 - 2t)$

Distributing: $4t - 8 + 2t - 4 = 18 - 4t$

Combining like terms: $6t - 12 = 18 - 4t$

Adding $4t$ to each side to move the variable: $10t - 12 = 18$

Adding 12 to each side to isolate the variable: $10t = 30$

Dividing each side by 10 to isolate the variable: $t = 3$

The answer can be checked by substituting the value for the variable into the original equation, ensuring that both sides calculate to be equal.

Linear inequalities express the relationship between unequal values. More specifically, they describe in what way the values are unequal. A value can be greater than (>), less than (<), greater than or equal to (≥), or less than or equal to (≤) another value. $5x + 40 > 65$ is read as *five times a number added to forty is greater than sixty-five.*

When solving a linear inequality, the solution is the set of all numbers that make the statement true. The inequality $x + 2 \geq 6$ has a solution set of 4 and every number greater than 4 (4.01; 5; 12; 107; etc.). Adding 2 to 4 or any number greater than 4 results in a value that is greater than or equal to 6. Therefore, $x \geq 4$ is the solution set.

To algebraically solve a linear inequality, follow the same steps as those for solving a linear equation. The inequality symbol stays the same for all operations *except* when multiplying or dividing by a negative number. If multiplying or dividing by a negative number while solving an inequality, the relationship reverses (the sign flips). In other words, > switches to < and vice versa. Multiplying or dividing by a positive number does not change the relationship, so the sign stays the same. An example is shown below.

Solve $-2x - 8 \leq 22$

Add 8 to both sides: $-2x \leq 30$

Divide both sides by -2: $x \geq -15$

Solutions of a linear equation or a linear inequality are the values of the variable that make a statement true. In the case of a linear equation, the solution set (list of all possible solutions) typically consists of a single numerical value. To find the solution, the equation is solved by isolating the variable. For example, solving the equation $3x - 7 = -13$ produces the solution $x = -2$. The only value for x which produces a true statement is -2. This can be checked by substituting -2 into the original equation to check that both sides are equal. In this case, $3(-2) - 7 = -13 \rightarrow -13 = -13$; therefore, -2 is a solution.

Although linear equations generally have one solution, this is not always the case. If there is no value for the variable that makes the statement true, there is no solution to the equation. Consider the equation $x + 3 = x - 1$. There is no value for x in which adding 3 to the value produces the same result as subtracting one from the value. Conversely, if any value for the variable makes a true statement, the equation has an infinite number of solutions. Consider the equation $3x + 6 = 3(x + 2)$. Any number substituted for x will result in a true statement (both sides of the equation are equal).

By manipulating equations like the two above, the variable of the equation will cancel out completely. If the remaining constants express a true statement (ex. $6 = 6$), then all real numbers are solutions to the equation. If the constants left express a false statement (ex. $3 = -1$), then no solution exists for the equation.

Solving a linear inequality requires all values that make the statement true to be determined. For example, solving $3x - 7 \geq -13$ produces the solution $x \geq -2$. This means that -2 and any number greater than -2 produces a true statement. Solution sets for linear inequalities will often be displayed using a number line. If a value is included in the set ($\geq$ or $\leq$), a shaded dot is placed on that value and an arrow extending in the direction of the solutions. For a variable > or $\geq$ a number, the arrow will point right on a number line, the direction where the numbers increase. If a variable is < or $\leq$ a number, the arrow will point left on a number line, which is the direction where the numbers decrease. If the value is not included in the set (> or <), an open (unshaded) circle on that value is used with an arrow in the appropriate direction.

It looks like this:

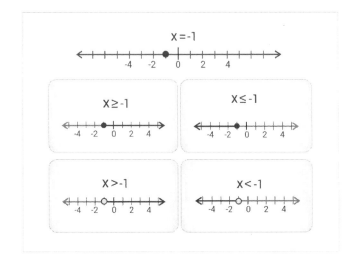

Similar to linear equations, a linear inequality may have a solution set consisting of all real numbers, or can contain no solution. When solved algebraically, a linear inequality in which the variable cancels out and results in a true statement (ex. $7 \geq 2$) has a solution set of all real numbers. A linear inequality in which the variable cancels out and results in a false statement (ex. $7 \leq 2$) has no solution.

Systems of Equations

To start, a review of linear equations is needed. When given a linear equation, the equation will show two expressions containing a variable that must be equal. Thus, one example would be $3x + 1 = 16$. To solve such an equation, remember two things. First, the final solution must equal x. Second, if two quantities are equal, one can add, subtract, multiply, or divide the same thing on both sides and end up with a true equation. In this case, subtract 1 from both sides, which would be a new equation, $3x = 15$. Then, divide both sides by 3 to get $x = 5$.

A system of equations can be solved by the same kinds of considerations, except that in this case, there are multiple equations that all have to be true at the same time. This means there are some new choices for finding solutions. First, if there are two equations, add the left side and the right side to get a new equation (the left side of the new equation is the sum of the left sides, and the right side of the new equation will be the sum of the right sides). Second, if one equation is solved in terms of one of the variables, the expression can be substituted into the other equation. Otherwise, the approach to solving these systems will be similar to solving a single equation.

A system of equations with at least one solution is called a *consistent system*. If a system has no solution, it is called an *inconsistent system*.

A *linear system* of equations with two variables and two equations is a system with variables x and y (or any other pair of variables) and equations that can be simplified to yield $ax + by = c, dx + ey = f$. There are two ways to solve such a system. The first is to solve for one variable in terms of the other and substitute it into the other equation. For example, from the first equation, $by = c - ax$, that means $y = \frac{c-ax}{b}$. $\frac{c-ax}{b}$ can be substituted for y in the second equation. This approach is called solving by *substitution*.

The other possibility is to multiply one of the equations on both sides by some constant, and then add the result to the other equation so that it eliminates one variable. For example, given the pair $ax + by = c, dx + ey = f$, multiply the first equation by $-\frac{d}{a}$. Then the first equation would become $-dx - \frac{db}{a}y = -\frac{cd}{a}$. Adding the equations results in the x terms cancelling, and an equation that only involves the variable y. This approach is called solving by *elimination*.

To illustrate the two approaches, use the system of equations: $2x + 4y = 6, x + y = 2$. This system will be solved using both methods.

By substitution: starting with the second equation, subtract y from both sides. The result of this step is $x = 2 - y$. Substitute $2 - y$ for x in the first equation, with a result of $2(2 - y) + 4y = 6$. This simplifies to $4 - 2y + 4y = 6, 2y = 2, y = 1$. Then, substitute 1 for y into $x = 2 - y$ to find the value for x: $x = 2 - 1 = 1$ or $x = 1$. So $x = 1, y = 1$.

To solve by elimination, starting with $2x + 4y = 6, x + y = 2$: to cancel the $2x$ in the first equation, place a -2x in the second equation on the left. The second equation is then multiplied by -2 on both sides, which gives $-2x - 2y = -4$. The equations are added together: $2x + 4y + (-2x - 2y) = 6 - 4$. The x terms cancel, and the result is $2y = 2$ or $y = 1$. Substituting this back into either of the original equations has a result of $x = 1$. So $x = 1, y = 1$.

Solving for X in Proportions

Proportions are commonly used to solve word problems to find unknown values such as x that are some percent or fraction of a known number. Proportions are solved by cross-multiplying and then dividing to arrive at x. The following examples show how this is done:

1) $\dfrac{75\%}{90\%} = \dfrac{25\%}{x}$

To solve for x, the fractions must be cross multiplied: $(75\%x = 90\% \times 25\%)$. To make things easier, let's convert the percentages to decimals: $(0.9 \times 0.25 = 0.225 = 0.75x)$. To get rid of x's co-efficient, each side must be divided by that same coefficient to get the answer $x = 0.3$. The question could ask for the answer as a percentage or fraction in lowest terms, which are 30% and $\frac{3}{10}$, respectively.

2) $\dfrac{x}{12} = \dfrac{30}{96}$

Cross-multiply: $96x = 30 \times 12$
Multiply: $96x = 360$
Divide: $x = 360 \div 96$
Answer: $x = 3.75$

3) $\dfrac{0.5}{3} = \dfrac{x}{6}$

Cross-multiply: $3x = 0.5 \times 6$
Multiply: $3x = 3$
Divide: $x = 3 \div 3$
Answer: $x = 1$

You may have noticed there's a faster way to arrive at the answer. If there is an obvious operation being performed on the proportion, the same operation can be used on the other side of the proportion to solve for x. For example, in the first practice problem, 75% became 25% when divided by 3, and upon doing the same to 90%, the correct answer of 30% would have been found with much less legwork. However, these questions aren't always so intuitive, so it's a good idea to work through the steps, even if the answer seems apparent from the outset.

Word Problems and Applications

In word problems, multiple quantities are often provided with a request to find some kind of relation between them. This often will mean that one variable (the dependent variable whose value needs to be found) can be written as a function of another variable (the independent variable whose value can be figured from the given information). The usual procedure for solving these problems is to start by giving each quantity in the problem a variable, and then figuring the relationship between these variables.

For example, suppose a car gets 25 miles per gallon. How far will the car travel if it uses 2.4 gallons of fuel? In this case, y would be the distance the car has traveled in miles, and x would be the amount of fuel burned in gallons (2.4). Then the relationship between these variables can be written as an algebraic equation, $y = 25x$. In this case, the equation is $y = 25 \cdot 2.4 = 60$, so the car has traveled 60 miles.

Some word problems require more than just one simple equation to be written and solved. Consider the following situations and the linear equations used to model them.

Suppose Margaret is 2 miles to the east of John at noon. Margaret walks to the east at 3 miles per hour. How far apart will they be at 3 p.m.? To solve this, x would represent the time in hours past noon, and y would represent the distance between Margaret and John. Now, noon corresponds to the equation where x is 0, so the y intercept is going to be 2. It's also known that the slope will be the rate at which the distance is changing, which is 3 miles per hour. This means that the slope will be 3 (be careful at this point: if units were used, other than miles and hours, for x and y variables, a conversion of the given information to the appropriate units would be required first). The simplest way to write an equation given the y-intercept, and the slope is the Slope-Intercept form, is $y = mx + b$. Recall that m here is the slope and b is the y intercept. So, $m = 3$ and $b = 2$. Therefore, the equation will be $y = 3x + 2$. The word problem asks how far to the east Margaret will be from John at 3 p.m., which means when x is 3. So, substitute $x = 3$ into this equation to obtain $y = 3 \cdot 3 + 2 = 9 + 2 = 11$. Therefore, she will be 11 miles to the east of him at 3 p.m.

For another example, suppose that a box with 4 cans in it weighs 6 lbs., while a box with 8 cans in it weighs 12 lbs. Find out how much a single can weighs. To do this, let x denote the number of cans in the box, and y denote the weight of the box with the cans in lbs. This line touches two pairs: $(4, 6)$ and $(8, 12)$. A formula for this relation could be written using the two-point form, with $x_1 = 4, y_1 = 6, x_2 = 8, y_2 = 12$. This would yield $\frac{y-6}{x-4} = \frac{12-6}{8-4}$, or $\frac{y-6}{x-4} = \frac{6}{4} = \frac{3}{2}$. However, only the slope is needed to solve this problem, since the slope will be the weight of a single can. From the computation, the slope is $\frac{3}{2}$. Therefore, each can weighs $\frac{3}{2}$ lb.

Practice Questions

1. $\dfrac{14}{15} + \dfrac{3}{5} - \dfrac{1}{30} =$

    a. $\dfrac{19}{15}$

    b. $\dfrac{43}{30}$

    c. $\dfrac{4}{3}$

    d. $\dfrac{3}{2}$

    e. 3

2. Solve for x and y, given $3x + 2y = 8, -x + 3y = 1$.
    a. $x = 2, y = 1$
    b. $x = 1, y = 2$
    c. $x = -1, y = 6$
    d. $x = 3, y = 1$
    e. $x = 4, y = 4$

3. $\dfrac{1}{2}\sqrt{16} =$
    a. 0
    b. 1
    c. 2
    d. 4
    e. 8

4. The factors of $2x^2 - 8$ are:
    a. $2(4x^2)$
    b. $2(x^2 + 4)$
    c. $2(x + 2)(x + 2)$
    d. $2(x - 2)(x - 2)$
    e. $2(x + 2)(x - 2)$

5. Two of the interior angles of a triangle are 35° and 70°. What is the measure of the last interior angle?
    a. 60°
    b. 75°
    c. 90°
    d. 100°
    e. 105°

6. A square field has an area of 400 square feet. What is its perimeter?
  a. 100 feet
  b. 80 feet
  c. $40\sqrt{2}$ feet
  d. 40 feet
  e. 2 feet

7. $\frac{5}{3} \times \frac{7}{6} =$

  a. $\frac{3}{5}$

  b. $\frac{18}{3}$

  c. $\frac{45}{31}$

  d. $\frac{17}{6}$

  e. $\frac{35}{18}$

8. One apple costs $2. One papaya costs $3. If Samantha spends $35 and gets 15 pieces of fruit, how many papayas did she buy?
  a. Three
  b. Four
  c. Five
  d. Six
  e. Seven

9. If $x^2 - 6 = 30$, then one possible value for x is:
  a. -6
  b. -4
  c. 3
  d. 5
  e. 8

10. A cube has a side length of 6 inches. What is its volume?
  a. 6 cubic inches
  b. 36 cubic inches
  c. 144 cubic inches
  d. 200 cubic inches
  e. 216 cubic inches

11. A square has a side length of 4 inches. A triangle has a base of 2 inches and a height of 8 inches. What is the total area of the square and triangle?
  a. 24 square inches
  b. 28 square inches
  c. 32 square inches
  d. 36 square inches
  e. 40 square inches

12. $-\frac{1}{3}\sqrt{81} =$
    a. -9
    b. -3
    c. 0
    d. 3
    e. 9

13. Simplify $(2x - 3)(4x + 2)$
    a. $8x^2 - 8x - 6$
    b. $6x^2 + 8x - 5$
    c. $-4x^2 - 8x - 1$
    d. $4x^2 - 4x - 6$
    e. $5x^2 + 4x + 3$

14. $\frac{11}{6} - \frac{3}{8} =$
    a. $\frac{5}{4}$
    b. $\frac{51}{36}$
    c. $\frac{35}{24}$
    d. $\frac{3}{2}$
    e. $\frac{39}{16}$

15. A triangle is to have a base $\frac{1}{3}$ as long as its height. Its area must be 6 square feet. How long will its base be?
    a. 1 foot
    b. 1.5 feet
    c. 2 feet
    d. 2.5 feet
    e. 3 feet

16. Which is closest to 17.8×9.9?
    a. 140
    b. 180
    c. 200
    d. 350
    e. 400

17. 6 is 30% of what number?
    a. 18
    b. 22
    c. 24
    d. 26
    e. 20

18. $3\frac{2}{3} - 1\frac{4}{5} =$

   a. $1\frac{13}{15}$

   b. $\frac{14}{15}$

   c. $2\frac{2}{3}$

   d. $\frac{4}{5}$

   e. $1\frac{2}{3}$

19. What is $\frac{420}{98}$ rounded to the nearest integer?
   a. 7
   b. 3
   c. 5
   d. 6
   e. 4

20. Which of the following is largest?
   a. 0.45
   b. 0.096
   c. 0.3
   d. 0.313
   e. 0.078

21. What is the value of b in this equation?

$$5b - 4 = 2b + 17$$

   a. 13
   b. 24
   c. 7
   d. 21
   e. 32

22. Twenty is 40% of what number?
   a. 50
   b. 8
   c. 200
   d. 5000
   e. 80

23. Which of the following expressions is equivalent to this equation?

$$\frac{2xy^2 + 4x - 8y}{16xy}$$

a. $\frac{y}{8} + \frac{1}{4y} - \frac{1}{2x}$

b. $8xy + 4y - 2x$

c. $xy^2 + \frac{x}{4y} - \frac{1}{2x}$

d. $\frac{y}{8} + 4y - 8y$

e. $\frac{y}{8} - 4y + 8y$

24. Arrange the following numbers from least to greatest value:

$0.85, \frac{4}{5}, \frac{2}{3}, \frac{91}{100}$

a. $0.85, \frac{4}{5}, \frac{2}{3}, \frac{91}{100}$

b. $\frac{4}{5}, 0.85, \frac{91}{100}, \frac{2}{3}$

c. $\frac{2}{3}, \frac{4}{5}, 0.85, \frac{91}{100}$

d. $0.85, \frac{91}{100}, \frac{4}{5}, \frac{2}{3}$

e. $\frac{2}{3}, 0.85, \frac{4}{5}, \frac{91}{10}$

25. Simplify the following expression:

$$(3x + 5)(x - 8)$$

a. $3x^2 - 19x - 40$

b. $4x - 19x - 13$

c. $3x^2 - 19x + 40$

d. $3x^2 + 5x - 3$

e. $4x - 5x + 9$

Answer Explanations

1. D: Start by taking a common denominator of 30. $\frac{14}{15} = \frac{28}{30}, \frac{3}{5} = \frac{18}{30}, \frac{1}{30} = \frac{1}{30}$. Add and subtract the numerators for the next step. $\frac{28}{30} + \frac{18}{30} - \frac{1}{30} = \frac{28+18-1}{30} = \frac{45}{30} = \frac{3}{2}$, where in the last step the 15 is factored out from the numerator and denominator.

2. A: From the second equation, add x to both sides and subtract 1 from both sides: $-x + 3y + x - 1 = 1 + x - 1$, with the result of $3y - 1 = x$. Substitute this into the first equation and get $3(3y - 1) + 2y = 8$, or $9y - 3 + 2y = 8, 11y = 11, y = 1$. Putting this into $3y - 1 = x$ gives $3(1) - 1 = x$ or $x = 2, y = 1$.

3. C: First, the square root of 16 is 4. So this simplifies to $\frac{1}{2}\sqrt{16} = \frac{1}{2}(4) = 2$.

4. E: The easiest way to approach this problem is to factor out a 2 from each term. $2x^2 - 8 = 2(x^2 - 4)$. Use the formula $x^2 - y^2 = (x + y)(x - y)$ to factor $x^2 - 4 = x^2 - 2^2 = (x + 2)(x - 2)$. So $2(x^2 - 4) = 2(x + 2)(x - 2)$.

5. B: The total of the interior angles of a triangle must be 180°. The sum of the first two is 105°, so the remaining is $180° - 105° = 75°$.

6. B: The length of the side will be $\sqrt{400}$. The calculation is performed a bit more easily by breaking this into the product of two square roots, $\sqrt{400} = \sqrt{4 \times 100} = \sqrt{4} \times \sqrt{100} = 2 \times 10 = 20$ feet. However, there are 4 sides, so the total is $20 \times 4 = 80$ feet.

7. E: To take the product of two fractions, just multiply the numerators and denominators. $\frac{5}{3} \times \frac{7}{6} = \frac{5 \times 7}{3 \times 6} = \frac{35}{18}$. The numerator and denominator have no common factors, so this is simplified completely.

8. C: Let a be the number of apples purchased, and let p be the number of papayas purchased. There is a total of 15 pieces of fruit, so one equation is $a + p = 15$. The total cost is \$35, and in terms of the total apples and papayas purchased as $2a + 3p = 35$. If we multiply the first equation by 2 on both sides, it becomes $2a + 2p = 30$. We then subtract this equation from the second equation: $2a + 3p - (2a + 2p) = 35 - 30, p = 5$. So five papayas were purchased.

9. A: This equation can be solved as follows: $x^2 = 36$, so $x = \pm\sqrt{36} = \pm 6$. Only -6 shows up in the list.

10. E: The volume of a cube is given by cubing the length of its side. $6^3 = 6 \times 6 \times 6 = 36 \times 6 = 216$.

11. A: The area of the square is the square of its side length, so $4^2 = 16$ square inches. The area of a triangle is half the base times the height, so $\frac{1}{2} \times 2 \times 8 = 8$ square inches. The total is $16 + 8 = 24$ square inches.

12. B: $-\frac{1}{3}\sqrt{81} = -\frac{1}{3}(9) = -3$

13. A: Multiply each of the terms in the first parentheses and then multiply each of the terms in the second parentheses. $(2x - 3)(4x + 2) = 2x(4x) + 2x(2) - 3(4x) - 3(2) = 8x^2 + 4x - 12x - 6 = 8x^2 - 8x - 6$.

14. C: Use a common denominator of 24. $\frac{11}{6} - \frac{3}{8} = \frac{44}{24} - \frac{9}{24} = \frac{44-9}{24} = \frac{35}{24}$.

15. C: The formula for the area of a triangle with base b and height h is $\frac{1}{2}bh$, where the base is one-third the height, or $b = \frac{1}{3}h$ or equivalently $h = 3b$. Using the formula for a triangle, this becomes $\frac{1}{2}b(3b) = \frac{3}{2}b^2$. Now, this has to be equal to 6. So $\frac{3}{2}b^2 = 6, b^2 = 4, b = \pm 2$. However, lengths are positive, so the base must be 2 feet long.

16. B: Instead of multiplying these out, the product can be estimated by using $18 \times 10 = 180$. The error here should be lower than 15, since it is rounded to the nearest integer, and the numbers add to something less than 30.

17. E: 30% is $\frac{3}{10}$. The number itself must be $\frac{10}{3}$ of 6, or $\frac{10}{3} \times 6 = 10 \times 2 = 20$.

18. A: These numbers to improper fractions: $\frac{11}{3} - \frac{9}{5}$. Take 15 as a common denominator: $\frac{11}{3} - \frac{9}{5} = \frac{55}{15} - \frac{27}{15} = \frac{28}{15} = 1\frac{13}{15}$ (when rewritten to get rid of the partial fraction).

19. E: Dividing by 98 can be approximated by dividing by 100, which would mean shifting the decimal point of the numerator to the left by 2. The result is 4.2 and rounds to 4.

20. A: To figure out which is largest, look at the first non-zero digits. Answer B and E's first nonzero digit is in the hundredths place. The other three all have nonzero digits in the tenths place, so it must be A, C, or D. Of these, A has the largest first nonzero digit.

21. C: To solve for the value of b, both sides of the equation need to be equalized.

Start by cancelling out the lower value of -4 by adding 4 to both sides:

$$5b - 4 = 2b + 17$$
$$5b - 4 + 4 = 2b + 17 + 4$$
$$5b = 2b + 21$$

The variable b is the same on each side, so subtract the lower 2b from each side:

$$5b = 2b + 21$$
$$5b - 2b = 2b + 21 - 2b$$
$$3b = 21$$

Then divide both sides by 3 to get the value of b:

$$3b = 21$$

$$\frac{3b}{3} = \frac{21}{3}$$

$$b = 7$$

22. A: Setting up a proportion is the easiest way to represent this situation. The proportion becomes $\frac{20}{x} = \frac{40}{100}$, where cross-multiplication can be used to solve for x. The answer can also be found by observing the two fractions as equivalent, knowing that twenty is half of forty, and fifty is half of one-hundred.

23. A: First, separate each element of the numerator with the denominator as follows:

$$\frac{2xy^2}{16xy} + \frac{4x}{16xy} - \frac{8y}{16xy}$$

Simplify each expression accordingly, reaching answer *A*:

$$\frac{y}{8} + \frac{1}{4y} - \frac{1}{2x}$$

24. C: The first step is to depict each number using decimals. $\frac{91}{100} = 0.91$

Multiplying both the numerator and denominator of $\frac{4}{5}$ by 20 makes it $\frac{80}{100}$ or 0.80; the closest approximation of $\frac{2}{3}$ would be $\frac{66}{100}$ or 0.66 recurring. Rearrange each expression in ascending order, as found in answer *C*.

25. A: When parentheses are around two expressions, they need to be *multiplied*. In this case, separate each expression into its parts (separated by addition and subtraction) and multiply by each of the parts in the other expression. Then, add the products together.

$$(3x)(x) + (3x)(-8) + (+5)(x) + (+5)(-8)$$

$$3x^2 - 24x + 5x - 40$$

Remember that when multiplying a positive integer by a negative integer, it will remain negative. Then add $-24x + 5x$ to get the simplified expression, answer *A*.

Self-Description Inventory

The self-description inventory is a personality test provided in the last section of the AFOQT. The self-description inventory is not included in the final score, so no one answer is better than another. Since these questions are helpful in assessing your personal characteristics, answer them as best you can and don't spend much time analyzing them. The purpose is to find an appropriate career match by comparing your answers to others who hold Air Force positions.

General Science

Physical Science

Structure of Matter

Elements, Compounds, and Mixtures

Everything that takes up space and has mass is composed of *matter*. Understanding the basic characteristics and properties of matter helps with classification and identification.

An *element* is a substance that cannot be chemically decomposed to a simpler substance, while still retaining the properties of the element.

Compounds are composed of two or more elements that are chemically combined. The constituent elements in the compound are in constant proportions by mass.

When a material can be separated by physicals means (such as sifting it through a colander), it is called a *mixture*. Mixtures are categorized into two types: *heterogeneous* and *homogeneous*. Heterogeneous mixtures have physically distinct parts, which retain their different properties. A mix of salt and sugar is an example of a heterogeneous mixture. With heterogenous mixtures, it is possible that different samples from the same parent mixture may have different proportions of each component in the mixture. For example, in the sugar and salt mixture, there may be uneven mixing of the two, causing one random tablespoon sample to be mostly salt, while a different tablespoon sample may be mostly sugar.

A homogeneous mixture, also called a *solution*, has uniform properties throughout a given sample. An example of a homogeneous solution is salt fully dissolved in warm water. In this case, any number of samples taken from the parent solution would be identical.

Atoms, Molecules, and Ions

The basic building blocks of matter are *atoms*, which are extremely small particles that retain their identity during chemical reactions. Atoms can be singular or grouped to form elements. Elements are composed of one type of atom with the same properties.

Molecules are a group of atoms—either the same or different types—that are chemically bonded together by attractive forces. For example, hydrogen and oxygen are both atoms but, when bonded together, form water.

Ions are electrically-charged particles that are formed from an atom or a group of atoms via the loss or gain of electrons.

Basic Properties of Solids, Liquids, and Gases

Matter exists in certain *states*, or physical forms, under different conditions. These states are called *solid*, *liquid*, or *gas*.

A solid has a rigid, or set, form and occupies a fixed shape and volume. Solids generally maintain their shape when exposed to outside forces.

Liquids and gases are considered fluids, which have no set shape. Liquids are fluid, yet are distinguished from gases by their incompressibility (incapable of being compressed) and set volume. Liquids can be transferred from one container to another, but cannot be forced to fill containers of different volumes via compression without causing damage to the container. For example, if one attempts to force a given volume or number of particles of a liquid, such as water, into a fixed container, such as a small water bottle, the container would likely explode from the extra water.

A gas can easily be compressed into a confined space, such as a tire or an air mattress. Gases have no fixed shape or volume. They can also be subjected to outside forces, and the number of gas molecules that can fill a certain volume vary with changes in temperature and pressure.

Basic Structure of an Atom

Atomic Models
Theories of the atomic model have developed over the centuries. The most commonly referenced model of an atom was proposed by Niels Bohr. Bohr studied the models of J.J. Thomson and Ernest Rutherford and adapted his own theories from these existing models. Bohr compared the structure of the atom to that of the Solar System, where there is a center, or nucleus, with various sized orbitals circulating around this nucleus. This is a simplified version of what scientists have discovered about atoms, including the structures and placements of any orbitals. Modern science has made further adaptations to the model, including the fact that orbitals are actually made of electron "clouds."

Atomic Structure: Nucleus, Electrons, Protons, and Neutrons

Following the Bohr model of the atom, the nucleus, or core, is made up of positively charged *protons* and neutrally charged *neutrons*. The neutrons are theorized to be in the nucleus with the protons to provide greater "balance" at the center of the atom. The nucleus of the atom makes up the majority (more than 99%) of the mass of an atom, while the orbitals surrounding the nucleus contain negatively charged *electrons*. The entire structure of an atom is incredibly small.

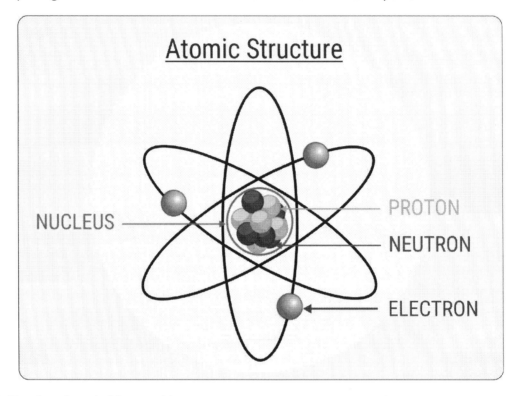

Atomic Number, Atomic Mass, and Isotopes

The *atomic number* of an atom is determined by the number of protons within the nucleus. When a substance is composed of atoms that all have the same atomic number, it is called an *element*. Elements are arranged by atomic number and grouped by properties in the *periodic table*.

An atom's *mass number* is determined by the sum of the total number of protons and neutrons in the atom. Most nuclei have a net neutral charge, and all atoms of one type have the same atomic number. However, there are some atoms of the same type that have a different mass number, due to an imbalance of neutrons. These are called *isotopes*. In isotopes, the atomic number, which is determined by the number of protons, is the same, but the mass number, which is determined by adding the protons and neutrons, is different due to the irregular number of neutrons.

Electron Arrangements

Electrons are most easily organized into distributions of subshells called *electron configurations*. Subshells fill from the inside (closest to the nucleus) to the outside. Therefore, once a subshell is filled, the next shell farther from the nucleus begins to fill, and so on. Atoms with electrons on the outside of a noble gas core (an atom with an electron inner shell that corresponds to the configuration of one of the noble gases, such as Neon) and pseudo-noble gas core (an atom with an electron inner shell that is similar to that of a noble gas core along with (n -1) d^{10} electrons), are called *valence* electrons. Valence electrons are primarily the electrons involved in chemical reactions. The similarities in their configurations account for similarities in properties of groups of elements. Essentially, the groups (vertical columns) on the periodic table all have similar characteristics, such as solubility and reactivity, due to their similar electron configurations.

Basic Characteristics of Radioactive Materials

Radioisotopes
As mentioned, an isotope is a variation of an element with a different number of neutrons in the nucleus, causing the nucleus to be unstable. When an element is unstable, it will go through decay or disintegration. All manmade elements are unstable and will break down. The length of time for an unstable element to break down is called the *half-life*. As an element breaks down, it forms other elements, known as daughters. Once a stable daughter is formed, the radioactive decay stops.

Characteristics of Alpha Particles, Beta Particles, and Gamma Radiation
As radioactive decay is occurring, the unstable element emits *alpha*, *beta*, and *gamma* radiation. Alpha and beta radiation are not as far-reaching or as powerful as gamma radiation. Alpha radiation is caused by the emission of two protons and two neutrons, while beta radiation is caused by the emission of either an electron or a positron. In contrast, gamma radiation is the release of photons of energy, not particles. This makes it the farthest-reaching and the most dangerous of these emissions.

Fission and Fusion
The splitting of an atom is referred to as fission, whereas the combination of two atoms into one is called fusion. To achieve fission and break apart an isotope, the unstable isotope is bombarded with high-speed particles. This process releases a large amount of energy and is what provides the energy in a nuclear power plant. Fusion occurs when two nuclei are merged to form a larger nucleus. The action of fusion also creates a tremendous amount of energy. To put the difference in the levels of energy between fission and fusion into perspective, the level of energy from fusion is what provides energy to the Earth's sun.

Basic Concepts and Relationships Involving Energy and Matter

The study of energy and matter, including heat and temperature, is called *thermodynamics*. There are four fundamental laws of thermodynamics, but the first two are the most commonly discussed.

First Law of Thermodynamics
The first law of thermodynamics is also known as the *conservation of energy*. This law states that energy cannot be created or destroyed, but is just transferred or converted into another form through a thermodynamic process. For example, if a liquid is boiled and then removed from the heat source, the liquid will eventually cool. This change in temperature is not because of a loss of energy or heat, but from a transfer of energy or heat to the surroundings. This can include the heating of nearby air molecules, or the transfer of heat from the liquid to the container or to the surface where the container is resting.

This law also applies to the idea of perpetual motion. A self-powered perpetual motion machine cannot exist. This is because the motion of the machine would inevitably lose some heat or energy to friction, whether from materials or from the air.

Second Law of Thermodynamics
The second law of thermodynamics is also known as the *law of entropy*. Entropy means chaos or disorder. In simple terms, this law means that all systems tend toward chaos. When one or more

systems interacts with another, the total entropy is the sum of the interacting systems, and this overall sum also tends toward entropy.

Conservation of Matter in Chemical Systems

The conservation of energy is seen in the conservation of matter in chemical systems. This is helpful when attempting to understand chemical processes, since these processes must balance out. This means that extra matter cannot be created or destroyed, it must all be accounted for through a chemical process.

Kinetic and Potential Energy

The conservation of energy also applies to the study of energy in physics. This is clearly demonstrated through the kinetic and potential energy involved in a system.

The energy of motion is called *kinetic energy*. If an object has height, or is raised above the ground, it has *potential energy*. The total energy of any given system is the sum of the potential energy and the kinetic energy of the subject (object) in the system.

Potential energy is expressed by the equation:

$PE = mgh$

Where m equals the object's mass, g equals acceleration caused by the gravitational force acting on the object, and h equals the height of the object above the ground.

Kinetic energy is expressed by the following equation:

$KE = \frac{1}{2} mv^2$

Where m is the mass of the object and v is the velocity of the object.

Conservation of energy allows the total energy for any situation to be calculated by the following equation:

$KE + PE$

For example, a roller coaster poised at the top of a hill has all potential energy, and when it reaches the bottom of that hill, as it is speeding through its lowest point, it has all kinetic energy. Halfway down the hill, the total energy of the roller coaster is about half potential energy and half kinetic energy. Therefore, the total energy is found by calculating both the potential energy and the kinetic energy and then adding them together.

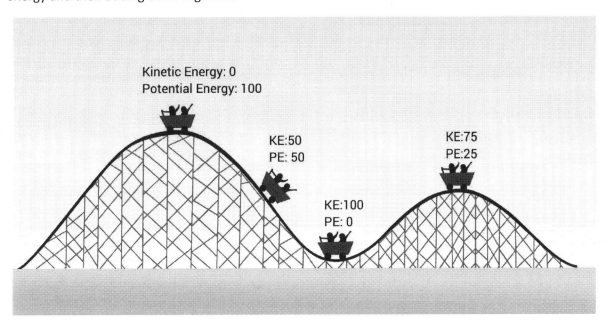

Transformations Between Different Forms of Energy

As stated by the conservation of energy, energy cannot be created or destroyed. If a system gains or loses energy, it is transformed within a single system from one type of energy to another or transferred from one system to another. For example, if the roller coaster system has potential energy that transfers to kinetic energy, the kinetic energy can then be transferred into thermal energy or heat released through braking as the coaster descends the hill. Energy can also transform from the chemical energy inside of a battery into the electrical energy that lights a train set. The energy released through nuclear fusion (when atoms are joined together, they release heat) is what supplies power plants with the energy for electricity. All energy is transferred from one form to another through different reactions. It can also be transferred through the simple action of atoms bumping into each other, causing a transfer of heat.

Differences Between Chemical and Physical Properties/Changes

A change in the physical form of matter, but not in its chemical identity, is known as a *physical change*. An example of a physical change is tearing a piece of paper in half. This changes the shape of the matter, but it is still paper.

Conversely, a *chemical change* alters the chemical composition or identity of matter. An example of a chemical change is burning a piece of paper. The heat necessary to burn the paper alters the chemical composition of the paper. This chemical change cannot be easily undone, since it has created at least one form of matter different than the original matter.

Temperature Scales

There are three main temperature scales used in science. The scale most often used in the United States is the *Fahrenheit* scale. This scale is based on the measurement of water freezing at 32° F and water boiling at 212° F. The Celsius scale uses 0° C as the temperature for water freezing and 100° C for water boiling. The accepted measurement by the International System of Units (from the French *Système international d'unités*), or SI, for temperature is the Kelvin scale. This is the scale employed in thermodynamics, since its zero is the basis for *absolute zero*, or the unattainable temperature, when matter no longer exhibits degradation.

The conversions between the temperature scales are as follows:

°Fahrenheit to °Celsius: $^0C = \frac{5}{9}(^0F - 32)$

°Celsius to °Fahrenheit: $^0F = \frac{9}{5}(^0C) + 32$

°Celsius to Kelvin: $K = \ ^0C + 273.15$

Transfer of Thermal Energy and Its Basic Measurement
There are three basic ways in which energy is transferred. The first is through *radiation*. Radiation is transmitted through electromagnetic waves and it does not need a medium to travel (it can travel in a vacuum). This is how the sun warms the Earth, and typically applies to large objects with great amounts of heat or objects with a large difference in their heat measurements.

The second form of heat transfer is *convection*. Convection involves the movement of "fluids" from one place to another. (The term *fluid* does not necessarily apply to a liquid, but any substance in which the molecules can slide past each other, such as gases.) It is this movement that transfers the heat to or from an area. Generally, convective heat transfer occurs through diffusion, which is when heat moves from areas of higher concentrations of particles to those of lower concentrations of particles and less heat. This process of flowing heat can be assisted or amplified through the use of fans and other methods of forcing the molecules to move.

The final process is called *conduction*. Conduction involves transferring heat through the touching of molecules. Molecules can either bump into each other to transfer heat, or they may already be touching each other and transfer the heat through this connection. For example, imagine a circular burner on an electric stove top. The coil begins to glow orange near the base of the burner that is connected to the stove because it heats up first. Since the burner is one continuous piece of metal, the molecules are touching each other. As they pass heat along the coil, it begins to glow all the way to the end.

To determine the amount of heat required to warm the coil in the above example, the type of material from which the coil is made must be known. The quantity of heat required to raise one gram of a substance one degree Celsius (or Kelvin) at a constant pressure is called *specific heat*. This measurement can be calculated for masses of varying substances by using the following equation:

$$q = s \times m \times \Delta t$$

Where *q* is the specific heat, *s* is the specific heat of the material being used, *m* is the mass of the substance being used, and *Δt* is the change in temperature.

A calorimeter is used to measure the heat of a reaction (either expelled or absorbed) and the temperature changes in a controlled system. A simple calorimeter can be made by using an insulated coffee cup with a thermometer inside. For this example, a lid of some sort would be preferred to prevent any escaping heat that could be lost by evaporation or convection.

Applications of Energy and Matter Relationships

When considering the cycling of matter in ecosystems, the flow of energy and atoms is from one organism to another. The *trophic level* of an organism refers to its position in a food chain. The level shows the relationship between it and other organisms on the same level and how they use and transfer energy to other levels in the food chain. This includes consumption and decomposition for the transfer of energy among organisms and matter. The sun provides energy through radiation to the Earth, and plants convert this light energy into chemical energy, which is then released to fuel the organism's activities.

Naturally occurring elements deep within the Earth's mantle release heat during their radioactive decay. This release of heat drives convection currents in the Earth's magma, which then drives plate tectonics. The transfer of heat from these actions causes the plates to move and create convection currents in the oceans. This type of cycling can also be seen in transformations of rocks. Sedimentary rocks can undergo significant amounts of heat and pressure to become metamorphic rocks. These rocks can melt back into magma, which then becomes igneous rock or, with extensive weathering and erosion, can revert to sediment and form sedimentary rocks over time. Under the right conditions (weathering and erosion), igneous rocks can also become sediment, which eventually compresses into sedimentary rock. Erosion helps the process by redepositing rocks into sediment on the sea floor.

All of these cycles are examples of the transfer of energy from one type into another, along with the conservation of mass from one level to the next.

Chemistry

<u>Periodicity and States of Matter</u>
Periodic Table of the Elements
Using the periodic table, elements are arranged by atomic number, similar characteristics, and electron configurations in a tabular format. The columns, called *groups*, are sorted by similar chemical properties and characteristics such as appearance and reactivity. This can be seen in the shiny texture of metals, the high melting points of alkali Earth metals, and the softness of post-transition metals. The rows are arranged by electron valance configurations and are called *periods*.

Periodic Table of the Elements

The elements are set in ascending order from left to right by atomic number. As mentioned, the atomic number is the number of protons contained within the nucleus of the atom. For example, the element helium has an atomic number of 2 because it has two protons in its nucleus.

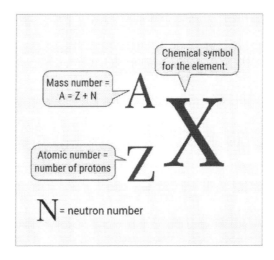

An element's mass number is calculated by adding the number of protons and neutrons of an atom together, while the atomic mass of an element is the weighted average of the naturally occurring atoms of a given element, or the relative abundance of isotopes that might be used in chemistry. For example, the atomic (mass) number of chlorine is 35; however, the atomic mass of chlorine is 35.5 amu (atomic mass unit). This discrepancy exists because there are many isotopes (meaning the nucleus could have 36 instead of 35 protons) occurring in nature. Given the prevalence of the various isotopes, the average of all of the atomic masses turns out to be 35.5 amu, which is slightly higher than chlorine's number on the periodic table. As another example, carbon has an atomic number of 12, but its atomic mass is 12.01 amu because, unlike chlorine, there are few naturally occurring isotopes to raise the average number.

Elements are arranged according to their valance electron configurations, which also contribute to trends in chemical properties. These properties help to further categorize the elements into blocks, including metals, non-metals, transition metals, alkali metals, alkali earth metals, metalloids, lanthanides, actinides, diatomics, post-transition metals, polyatomic non-metals, and noble gases. Noble gases (the far-right column) have a full outer electron valence shell. The elements in this block possess similar characteristics such as being colorless, odorless, and having low chemical reactivity. Another block, the metals, tend to be shiny, highly conductive, and easily form alloys with each other, non-metals, and noble gases.

The symbols of the elements on the periodic table are a single letter or a two-letter combination that is usually derived from the element's name. Many of the elements have Latin origins for their names, and their atomic symbols do not match their modern names. For example, iron is derived from the word *ferrum*, so its symbol is Fe, even though it is now called iron. The naming of the elements began with those of natural origin and their ancient names, which included the use of the ending "ium." This naming practice has been continued for all elements that have been named since the 1940s. Now, the names of new elements must be approved by the International Union of Pure and Applied Chemistry.

The elements on the periodic table are arranged by number and grouped by trends in their physical properties and electron configurations. Certain trends are easily described by the arrangement of the periodic table, which includes the increase of the atomic radius as elements go from right to left and from top to bottom on the periodic table. Another trend on the periodic table is the increase in ionization energy (or the tendency of an atom to attract and form bonds with electrons). This tendency increases from left to right and from bottom to top of the periodic table—the opposite directions of the trend for the atomic radius. The elements on the right side and near the bottom of the periodic table tend to attract electrons with the intent to gain, while the elements on the left and near the top usually lose, or give up, one or more electrons in order to bond. The only exceptions to this rule are the noble gases. Since the noble gases have full valence shells, they do not have a tendency to lose or gain electrons.

Chemical reactivity is another trend identifiable by the groupings of the elements on the periodic table. The chemical reactivity of metals decreases from left to right and while going higher on the table. Conversely, non-metals increase in chemical reactivity from left to right and while going lower on the table. Again, the noble gases present an exception to these trends because they have very low chemical reactivity.

Trends in the Periodic Table

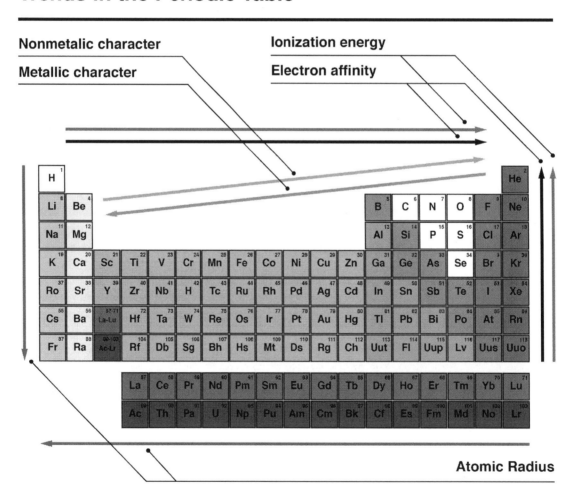

States of Matter and Factors that Affect Phase Changes

Matter is most commonly found in three distinct states: solid, liquid, and gas. A solid has a distinct shape and a defined volume. A liquid has a more loosely defined shape and a definite volume, while a gas has no definite shape or volume. The *Kinetic Theory of Matter* states that matter is composed of a large number of small particles (specifically, atoms and molecules) that are in constant motion. The distance between the separations in these particles determines the state of the matter: solid, liquid, or gas. In gases, the particles have a large separation and no attractive forces. In liquids, there is moderate separation between particles and some attractive forces to form a loose shape. Solids have almost no separation between their particles, causing a defined and set shape. The constant movement of particles causes them to bump into each other, thus allowing the particles to transfer energy between each other. This bumping and transferring of energy helps explain the transfer of heat and the relationship between pressure, volume, and temperature.

The *Ideal Gas Law* states that pressure, volume, and temperature are all related through the equation: $PV = nRT$, where P is pressure, V is volume, n is the amount of the substance in moles, R is the gas constant, and T is temperature.

Through this relationship, volume and pressure are both proportional to temperature, but pressure is inversely proportional to volume. Therefore, if the equation is balanced, and the volume decreases in the system, pressure needs to proportionately increase to keep both sides of the equation balanced. In contrast, if the equation is unbalanced and the pressure increases, then the temperature would also increase, since pressure and temperature are directly proportional.

When pressure, temperature, or volume change in matter, a change in state can occur. Changes in state include solid to liquid (melting), liquid to gas (evaporation), solid to gas (sublimation), gas to solid (deposition), gas to liquid (condensation), and liquid to solid (freezing). There is one other state of matter called *plasma*, which is seen in lightning, television screens, and neon lights. Plasma is most commonly converted from the gas state at extremely high temperatures.

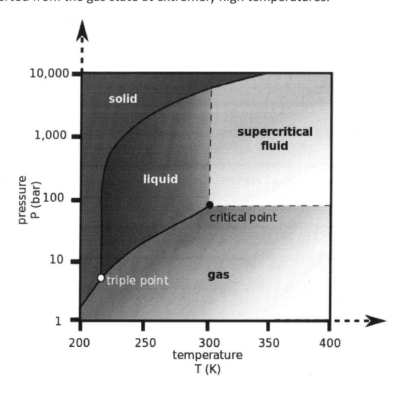

The amount of energy needed to change matter from one state to another is labeled by the terms for phase changes. For example, the temperature needed to supply enough energy for matter to change from a liquid to a gas is called the *heat of vaporization*. When heat is added to matter in order to cause a change in state, there will be an increase in temperature until the matter is about to change its state. During its transition, all of the added heat is used by the matter to change its state, so there is no increase in temperature. Once the transition is complete, then the added heat will again yield an increase in temperature.

Each state of matter is considered to be a phase, and changes between phases are represented by phase diagrams. These diagrams show the effects of changes in pressure and temperature on matter. The states of matter fall into areas on these charts called *heating curves*.

Chemical Nomenclature, Composition, and Bonding
Simple Compounds and Their Chemical Formulas
Chemical formulas represent the proportion of the number of atoms in a chemical compound. Chemical symbols are used for the elements present and numerical values. Parentheses are also sometimes used to show the number of combinations of the elements in relation to their ionic charges. An element's ionic charge can be determined by its location on the periodic table. This information is then used to correctly combine its atoms in a compound.

For example, the chemical formula for sodium chloride (table salt) is the combination of sodium (Na, ionic charge of +1) and chlorine (Cl, ionic charge of -1). From its placement on the periodic table, the electron valence of an outer shell can be determined: sodium has an ionic charge of +1, while chlorine has an ionic charge of -1. Since these two elements have an equal and opposite amount of charge, they combine in a neutral one-to-one ratio: NaCl. The naming of compounds depends mainly on the second element in a chemical compound. If it is a non-metal (such as chlorine), it is written with an "ide" at the end. The compound NaCl is called "sodium chloride."

If the elements forming a compound do not have equal and opposite ionic charges, there will be an unequal number of each element in the compound to balance the ionic charge. This situation happens with many elements, for example, in the combination of nickel and oxygen into nickel oxide (Ni_2O_3). Nickel has a +3 ionic charge and oxygen has a -2 ionic charge, so when forming a compound, there must be two nickel atoms for every three oxygen atoms (a common factor of 6) to balance the charge of the compound. This compound is called "nickel oxide."

A chemical formula can also be written from a compound's name. For instance, the compound carbon dioxide is formed by the combination of carbon and oxygen. The word "dioxide" means there are two oxygen atoms for every carbon atom, so it is written as CO_2.

To better represent the composition of compounds, structural formulas are used. The combination of atoms is more precisely depicted by lining up the electron configuration of the outer electron shell through a Lewis dot diagram.

The Lewis dot diagram, named for Gilbert N. Lewis, shows the arrangement of the electrons in the outer shell and how these electrons can pair/bond with the outer shell electrons of other atoms when forming compounds. The diagram is created by writing the symbol of an element and then drawing dots to represent the outer shell of valence electrons around what would be an invisible square surrounding the symbol. The placement of the first two dots can vary based on the school of teaching. For the given example, the first dot is placed on the top and then the next dot is placed beside it, since it represents the pair of electrons in the 1s valence shell. The next dots (electrons) are placed one at a time on each side—right, bottom, left, right bottom left, etc.—of the element symbol until all of the valence shell electrons are represented, or the structure has eight dots (electrons), which means it is full. This method gives a more specific picture of compounds, how they are structured, and what electrons are available for bonding, sharing, and forming new compounds. For example, the compound sodium chloride is written separately with sodium having one valence electron and chlorine having seven valence electrons. Then, combined with a total of eight electrons, it is written with two dots being shared between the two elements.

Lewis structure NaCl

$$Na\cdot \; + \; \cdot\ddot{C}l\!: \; \longrightarrow \; Na^+ + \; \ddot{\underset{..}{:}Cl}\!:$$

Types of Chemical Bonding

A chemical bond is a strong attractive force that can exist between atoms. The bonding of atoms is separated into two main categories. The first category, *ionic bonding,* primarily describes the bonding that occurs between oppositely charged ions in a regular crystal arrangement. It primarily exists between salts, which are known to be ionic. Ionic bonds are held together by the electrostatic attraction between oppositely charged ions. This type of bonding involves the transfer of electrons from the valence shell of one atom to the valence shell of another atom. If an atom loses an electron from its valence shell, it becomes a positive ion, or *cation*. If an atom gains an electron, it becomes a negative ion, or an *anion*. The Lewis electron-dot symbol is used to more simply express the electron configuration of atoms, especially when forming bonds.

The second type of bonding is covalent bonding. This bonding involves the sharing of a pair of electrons between atoms. There are no ions involved in covalent bonding, but the force holding the atoms together comes from the balance between the attractive and repulsive forces involving the shared electron and the nuclei. Atoms frequently engage is this type of bonding when it enables them to fill their outer valence shell.

Mole Concept and Its Applications

The calculation of mole ratios of reactants and products involved in a chemical reaction is called "stoichiometry." To find these ratios, one must first find the proportion of the number of molecules in one mole of a substance. This relates the molar mass of a compound to its mass and this relationship is a constant known as *Avogadro's number* (6.23×10^{23}). Since it is a ratio, there are no dimensions (or units) for Avogadro's number.

Molar Mass and Percent Composition

The molar mass of a substance is the measure of the mass of one mole of the substance. For pure elements, the molar mass is also known as the atomic mass unit (amu) of the substance. For compounds, it can be calculated by adding the molar masses of each substance in the compound. For example, the molar mass of carbon is 12.01 g/mol, while the molar mass of water (H_2O) requires finding the sum of the molar masses of the constituents ((1.01 x 2 = 2.02 g/mol for hydrogen) + (16.0 g/mol for oxygen) = 18.02 g/mol).

The percentage of a compound in a composition can be determined by taking the individual molar masses of each component divided by the total molar mass of the compound, multiplied by 100. Determining the percent composition of carbon dioxide (CO_2) first requires the calculation of the molar mass of CO_2.

molar mass of carbon = 12.01 x 1 atom = 12.01 g/mol

molar mass of oxygen = 16.0 × 2 atoms = 32.0 g/mol

molar mass of CO_2 = 12.01 g/mol + 32.0 g/mol = 44.01 g/mol

Next, each individual mass is divided by the total mass and multiplied by 100 to get the percent composition of each component.

12.01/44.01 = (0.2729 × 100) = 27.29% carbon

32.0/44.01 = (0.7271 × 100) = 72.71% oxygen

(A quick check in the addition of the percentages should always yield 100%.)

Chemical Reactions
Basic Concepts of Chemical Reactions
Chemical reactions rearrange the initial atoms of the reactants into different substances. These types of reactions can be expressed through the use of balanced chemical equations. A *chemical equation* is the symbolic representation of a chemical reaction through the use of chemical terms. The reactants at the beginning (or on the left side) of the equation must equal the products at the end (or on the right side) of the equation.

For example, table salt (NaCl) forms through the chemical reaction between sodium (Na) and chlorine (Cl) and is written as: Na + Cl_2 → NaCl.

However, this equation is not balanced because there are two sodium atoms for every pair of chlorine atoms involved in this reaction. So, the left side is written as: 2Na + Cl_2 → NaCl.

Next, the right side needs to balance the same number of sodium and chlorine atoms. So, the right side is written as: 2Na + Cl_2 → 2NaCl. Now, this is a balanced chemical equation.

Chemical reactions typically fall into two types of categories: *endothermic* and *exothermic*.

An endothermic reaction absorbs heat, whereas an exothermic reaction releases heat. For example, in an endothermic reaction, heat is drawn from the container holding the chemicals, which cools the container. Conversely, an exothermic reaction emits heat from the reaction and warms the container holding the chemicals.

Factors that can affect the rate of a reaction include temperature, pressure, the physical state of the reactants (e.g., surface area), concentration, and catalysts/enzymes.

The formula *PV = nRT* shows that an increase in any of the variables (pressure, volume, or temperature) affects the overall reaction. The physical state of two reactants can also determine how much interaction they have with each other. If two reactants are both in a fluid state, they may have the capability of interacting more than if solid. The addition of a catalyst or an enzyme can increase the rate of a chemical reaction, without the catalyst or enzyme undergoing a change itself.

Le Chatelier's principle describes factors that affect a reaction's equilibrium. Essentially, when introducing a "shock" to a system (or chemical reaction), a positive feedback/shift in equilibrium is often the response. In accordance with the second law of thermodynamics, this imbalance will eventually even itself out, but not without counteracting the effects of the reaction.

There are many different types of chemical reactions. A *synthesis reaction* is the combination of two or more elements into a compound. For example, the synthesis reaction of hydrogen and oxygen forms water.

$$2 H_2(g) + O_2(g) \rightarrow 2 H_2O(g)$$

A *decomposition reaction* is the breaking down of a compound into its more basic components. For example, the decomposition, or electrolysis, of water results in it breaking down into oxygen and hydrogen gas.

$$2 H_2O \rightarrow 2 H_2 + O_2$$

A *combustion reaction* is similar to a decomposition reaction, but it requires oxygen and heat for the reaction to occur. For example, the burning of a candle requires oxygen to ignite and the reaction forms carbon dioxide during the process.

$$CH_4(g) + 2O_2(g) \rightarrow CO_2(g) + 2H_2O(g)$$

There are also single and double replacement reactions where compounds swap components with each other to form new compounds. In the *single replacement reaction*, a single element will swap into a compound, thus releasing one of the compound's elements to become the new single element. For example, the reaction between iron and copper sulfate will create copper and iron sulfate.

$$1Fe(s) + 1CuSO_4(aq) \rightarrow 1FeSO_4(aq) + 1Cu(s)$$

In a *double replacement reaction*, two compounds swap components to form two new compounds. For example, the reaction between sodium sulfide and hydrochloric acid forms sodium chloride and hydrogen sulfide.

$$Na_2S + HCl \rightarrow NaCl + H_2S$$

An organic reaction is a chemical reaction involving the components of carbon and hydrogen.

Finally, there are oxidation/reduction (redox or half) reactions. These reactions involve the loss of electrons from one species (oxidation), and the gain of electrons to the other species (reduction). For example, the oxidation of magnesium is as follows:

$$2 Mg(s) + O_2(g) \rightarrow 2 MgO(s)$$

Acid-Base Chemistry
Simple Acid-Base Chemistry
If something has a sour taste, it is acidic, and if something has a bitter taste, it is basic. Unfortunately, it can be extremely dangerous to ingest chemicals in an attempt to classify them as an acid or a base. Therefore, acids and bases are generally identified by the reactions they have when combined with water. An acid will increase the concentration of the hydrogen ion (H^+), while a base will increase the concentration of the hydroxide ion (OH^-).

To better categorize the varying strengths of acids and bases, the pH scale is used. The pH scale provides a logarithmic (base 10) grading to acids and bases based on their strength. The pH scale contains values from 0 through 14, with 7 being neutral. If a solution registers below 7 on the pH scale, it is considered an acid. If it registers higher than 7, it is considered a base. To perform a quick test on a solution, litmus paper can be used. A base will turn red litmus paper blue, whereas an acid will turn blue litmus paper red. To gauge the strength of an acid or base, a test of phenolphthalein can be used. An acid will turn red phenolphthalein colorless, and a base will turn colorless phenolphthalein pink. As demonstrated with these types of tests, acids and bases neutralize each other. When acids and bases react with one another, they produce salts (also called ionic substances).

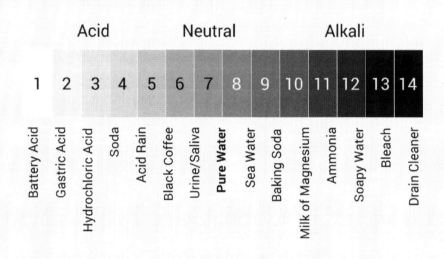

Solutions and Solubility

Different Types of Solutions

A *solution* is a homogenous mixture of more than one substance. A *solute* is another substance that can be dissolved into a substance called a *solvent*. If only a small amount of solute is dissolved in a solvent, the solution formed is said to be *diluted*. If a large amount of solute is dissolved into the solvent, then the solution is said to be *concentrated*. For example, water from a typical, unfiltered household tap is diluted because it contains other minerals in very small amounts.

Solution Concentration

A

Dilute solution

B

Concentrated solution

If more solute is being added to a solvent, but not dissolving, the solution is called *saturated*. For example, when hummingbirds eat sugar-water from feeders, they prefer it as sweet as possible. When trying to dissolve enough sugar (solute) into the water (solvent), there will be a point where the sugar crystals will no longer dissolve into the solution and will remain as whole pieces floating in the water. At this point, the solution is considered saturated and cannot accept more sugar. This level, at which a solvent cannot accept and dissolve any more solute, is called its *saturation point*. In some cases, it is possible to force more solute to be dissolved into a solvent, but this will result in crystallization. The state of a solution on the verge of crystallization, or in the process of crystallization, is called a *supersaturated* solution. This can also occur in a solution that seems stable, but if it is disturbed, the change can begin the crystallization process.

Although the terms *dilute*, *concentrated*, *saturated*, and *supersaturated* give qualitative descriptions of solutions, a more precise quantitative description needs to be established for the use of chemicals. This holds true especially for mixing strong acids or bases. The method for calculating the concentration of a solution is done through finding its molarity. In some instances, such as environmental reporting, molarity is measured in parts per million (ppm). Parts per million, is the number of milligrams of a substance dissolved in one liter of water. To find the *molarity*, or the amount of solute per unit volume of solution, for a solution, the following formula is used:

$$c = \frac{n}{V}$$

In this formula, c is the molarity (or unit moles of solute per volume of solution), n is the amount of solute measured in moles, and V is the volume of the solution, measured in liters.

Example:

What is the molarity of a solution made by dissolving 2.0 grams of NaCl into enough water to make 100 mL of solution?

To solve this, the number of moles of NaCl needs to be calculated:

First, to find the mass of NaCl, the mass of each of the molecule's atoms is added together as follows:

$$23.0g \text{ (Na)} + 35.5g \text{ (Cl)} = 58.8g \text{ NaCl}$$

Next, the given mass of the substance is multiplied by one mole per total mass of the substance:

$$2.0g \text{ NaCl} \times (1 \text{ mol NaCl}/58.5g \text{ NaCl}) = 0.034 \text{ mol NaCl}$$

Finally, the moles are divided by the number of liters of the solution to find the molarity:

$$(0.034 \text{ mol NaCl})/(0.100L) = 0.34 \text{ M NaCl}$$

To prepare a solution of a different concentration, the *mass solute* must be calculated from the molarity of the solution. This is done via the following process:

Example:

How would you prepare 600.0 mL of 1.20 M solution of sodium chloride?

To solve this, the given information needs to be set up:

$$1.20 \text{ M NaCl} = 1.20 \text{ mol NaCl}/1.00 \text{ L of solution}$$

$$0.600 \text{ L solution} \times (1.20 \text{ mol NaCl}/1.00 \text{ L of solution}) = 0.72 \text{ moles NaCl}$$

$$0.72 \text{ moles NaCl} \times (58.5g \text{ NaCl}/1 \text{ mol NaCl}) = 42.12 \text{ g NaCl}$$

This means that one must dissolve 42.12 g NaCl in enough water to make 600.0 L of solution.

Factors Affecting the Solubility of Substances and the Dissolving Process
Certain factors can affect the rate in dissolving processes. These include temperature, pressure, particle size, and agitation (stirring). As mentioned, the *ideal gas law* states that $PV = nRT$, where P equals pressure, V equals volume, and T equals temperature. If the pressure, volume, or temperature are affected in a system, it will affect the entire system. Specifically, if there is an increase in temperature, there will be an increase in the dissolving rate. An increase in the pressure can also increase the dissolving rate. Particle size and agitation can also influence the dissolving rate, since all of these factors contribute to the breaking of intermolecular forces that hold solute particles together. Once these forces are broken, the solute particles can link to particles in the solvent, thus dissolving the solute.

A *solubility curve* shows the relationship between the mass of solute that a solvent holds at a given temperature. If a reading is on the solubility curve, the solvent is *full* (*saturated*) and cannot hold anymore solute. If a reading is above the curve, the solvent is *unstable* (*supersaturated*) from holding more solute than it should. If a reading is below the curve, the solvent is *unsaturated* and could hold more solute.

If a solvent has different electronegativities, or partial charges, it is considered to be *polar*. Water is an example of a polar solvent. If a solvent has similar electronegativities, or lacking partial charges, it is considered to be *non-polar*. Benzene is an example of a non-polar solvent. Polarity status is important when attempting to dissolve solutes. The phrase "like dissolves like" is the key to remembering what will happen when attempting to dissolve a solute in a solvent. A polar solute will dissolve in a like, or polar solvent. Similarly, a non-polar solute will dissolve in a non-polar solvent. When a reaction produces a solid, the solid is called a *precipitate.* A precipitation reaction can be used for removing a salt (an ionic compound that results from a neutralization reaction) from a solvent, such as water. For water, this process is called *ionization*. Therefore, the products of a neutralization reaction (when an acid and base react) are a salt and water.

When a solute is added to a solvent to lower the freezing point of the solvent, it is called *freezing point depression*. This is a useful process, especially when applied in colder temperatures. For example, the addition of salt to ice in winter allows the ice to melt at a much lower temperature, thus creating safer road conditions for driving. Unfortunately, the freezing point depression from salt can only lower the melting point of ice so far, and is ineffectual when temperatures are too low. This same process, with a mix of ethylene glycol and water, is also used to keep the radiator fluid (antifreeze) in an automobile from freezing during the winter.

Physics

Mechanics
Description of Motion in One and Two Dimensions
The description of motion is known as *kinetics*, and the causes of motion are known as *dynamics*. Motion in one dimension is known as a *scalar* quantity. It consists of one measurement such as length (length or distance is also known as displacement), speed, or time. Motion in two dimensions is known as a *vector* quantity. This would be a speed with a direction, or velocity.

Velocity is the measure of the change in distance over the change in time. All vector quantities have a direction that can be relayed through the sign of an answer, such as -5.0 m/s or +5.0 m/s. The objects registering these velocities would be in opposite directions, where the change in distance is denoted by Δx and the change in time is denoted by Δt:

$$v = \frac{\Delta x}{\Delta t}$$

Acceleration is the measure of the change in an object's velocity over a change in time, where the change in velocity, $v_2 - v_1$, is denoted by Δv and the change in time, $t_1 - t_2$, is denoted by Δt:

$$a = \frac{\Delta v}{\Delta t}$$

The linear momentum, *p*, of an object is the result of the objects mass, *m*, multiplied by its velocity, *v*, and is described by the equation:

$$p = mv$$

This aspect becomes important when one object hits another object. For example, the linear momentum of a small sports car will be much smaller than the linear momentum of a large semi-truck. Thus, the semi-truck will cause more damage to the car than the car to the truck.

Newton's Three Laws of Motion
Sir Isaac Newton summarized his observations and calculations relating to motion into three concise laws.

First Law of Motion: Inertia
This law states that an object in motion tends to stay in motion or an object at rest tends to stay at rest, unless the object is acted upon by an outside force.

For example, a rock sitting on the ground will remain in the same place, unless it is pushed or lifted from its place.

The First Law also includes the relation of weight to gravity and force between objects relative to the distance separating them.

$$Weight = G\frac{Mm}{r^2}$$

In this equation, G is the gravitational constant, M and m are the masses of the two objects, and r is the distance separating the two objects.

Second Law of Motion: F = ma
This law states that the force on a given body is the result of the object's mass multiplied by any acceleration acting upon the object. For objects falling on Earth, an acceleration is caused by gravitational force (9.8 m/s^2).

Third Law of Motion: Action-Reaction
This law states that for every action there is an equal and opposite reaction. For example, if a person punches a wall, the wall exerts a force back on the person's hand equal and opposite to his or her punching force. Since the wall has more mass, it absorbs the impact of the punch better than the person's hand.

Mass, Weight, and Gravity
Mass is a measure of how much of a substance exists, or how much inertia an object has. The mass of an object does not change based on the object's location, but the weight of an object does vary with its location.

For example, a 15-kg mass has a weight that is determined by acceleration from the force of gravity here on Earth. However, if that same 15-kg mass were to be weighed on the moon, it would weigh much less, since the acceleration force from the moon's gravity is approximately one-sixth of that on Earth.

Weight = mass × acceleration

$$W_{Earth} = 15 \text{ kg} \times 9.8 \text{ m/s}^2 \qquad > \qquad W_{Moon} = 15 \text{ kg} \times 1.62 \text{ m/s}^2$$

$$W_{Earth} = 147N \qquad > \qquad 24.3N$$

Analysis of Motion and Forces

Projectile Motion describes the path of an object in the air. Generally, it is described by two-dimensional movement, such as a stone thrown through the air. This activity maps to a parabolic curve. However, the definition of projectile motion also applies to free fall, or the non-arced motion of an object in a path straight up and/or straight down. When an object is thrown horizontally, it is subject to the same influence of gravity as an object that is dropped straight down. The farther the projectile motion, the farther the distance of the object's flight.

Friction is a force that opposes motion. It can be caused by a number of materials; there is even friction caused by air. Whenever two differing materials touch, rub, or pass by each other, it will create friction, or an oppositional force, unless the interaction occurs in a true vacuum. To move an object across a floor, the force exerted on the object must overcome the frictional force keeping the object in place. Friction is also why people can walk on surfaces. Without the oppositional force of friction to a shoe pressing on the floor, a person would not be able to grip the floor to walk—similar to the challenge of walking on ice. Without friction, shoes slip and are unable to help people propel forward and walk.

When calculating the effects of objects hitting (or colliding with) each other, several things are important to remember. One of these is the definition of momentum: the mass of an object multiplied by the object's velocity. As mentioned, it is expressed by the following equation:

$$p = mv$$

Here, *p* is equal to an object's momentum, *m* is equal to the object's mass, and *v* is equal to the object's velocity.

Another important thing to remember is the principle of the conservation of linear momentum. The total momentum for objects in a situation will be the same before and after a collision. There are two primary types of collisions: elastic and inelastic. In an elastic collision, the objects collide and then travel in different directions. During an inelastic collision, the objects collide and then stick together in their final direction of travel. The total momentum in an elastic collision is calculated by using the following formula:

$$m_1 v_1 + m_2 v_2 = m_1 v_1 + m_2 v_2$$

Here, m_1 and m_2 are the masses of two separate objects, and v_1 and v_2 are the velocities, respectively, of the two separate objects.

The total momentum in an inelastic collision is calculated by using the following formula:

$$m_1 v_1 + m_2 v_2 = (m_1 + m_2) v_f$$

Here, v_f is the final velocity of the two masses after they stick together post-collision.

Example:

If two bumper cars are speeding toward each other, head-on, and collide, they are designed to bounce off of each other and head in different directions. This would be an elastic collision.

If real cars are speeding toward each other, head-on, and collide, there is a good chance their bumpers might get caught together and their direction of travel would be together in the same direction.

An *axis* is an invisible line on which an object can rotate. This is most easily observed with a toy top. There is actually a point (or rod) through the center of the top on which the top can be observed to be spinning. This is called the axis.

When objects move in a circle by spinning on their own axis, or because they are tethered around a central point (also an axis), they exhibit circular motion. Circular motion is similar in many ways to linear (straight line) motion; however, there are a few additional points to note. A spinning object is always accelerating because it is always changing direction. The force causing this constant acceleration on or around an axis is called *centripetal force* and is often associated with centripetal acceleration. Centripetal force always pulls toward the axis of rotation. An imaginary reactionary force, called *centrifugal force*, is the outward force felt when an object is undergoing circular motion. This reactionary force is not the real force; it just feels like it is there. For this reason, it has also been referred to as a "fictional force." The true force is the one pulling inward, or the centripetal force.

The terms *centripetal* and *centrifugal* are often mistakenly interchanged. If the centripetal force acting on an object moving with circular motion is removed, the object will continue moving in a straight line tangent to the point on the circle where the object last experienced the centripetal force. For example, when a traditional style washing machine spins a load of clothes to expunge the water from the load, it rapidly spins the machine barrel. A force is pulling in toward the center of the circle (centripetal force). At the same time, the wet clothes, which are attempting to move in a straight line, are colliding with the outer wall of the barrel that is moving in a circle. The interaction between the wet clothes and barrel wall cause a reactionary force to the centripetal force and this expels the water out of the small holes that line the outer wall of the barrel.

Conservation of Angular Momentum
An object moving in a circular motion also has momentum; for circular motion, it is called *angular momentum*. This is determined by rotational inertia, rotational velocity, and the distance of the mass from the axis or center of rotation. When objects exhibit circular motion, they also demonstrate the *conservation of angular momentum*, meaning that the angular momentum of a system is always constant, regardless of the placement of the mass. Rotational inertia can be affected by how far the mass of the object is placed with respect to the axis of rotation. The greater the distance between the mass and the axis of rotation, the slower the rotational velocity. Conversely, if the mass is closer to the axis of rotation, the rotational velocity is faster. A change in one affects the other, thus conserving the angular momentum. This holds true as long as no external forces act upon the system.

For example, ice skaters spinning in on one ice skate extends their arms out for a slower rotational velocity. When skaters bring their arms in close to their bodies (which lessens the distance between the mass and the axis of rotation), their rotational velocity increases and they spin much faster. Some skaters extend their arms straight up above their head, which causes an extension of the axis of rotation, thus removing any distance between the mass and the center of rotation, which maximizes their rotational velocity.

Another example is when a person selects a horse on a merry-go-round: the placement of their horse can affect their ride experience. All of the horses are traveling with the same rotational speed, but in order to travel along the same plane as the merry-go-round turns, a horse on the outside will have a greater linear speed because it is further away from the axis of rotation. Essentially, an outer horse has to cover a lot more ground than a horse on the inside in order to keep up with the rotational speed of the merry-go-round platform. Thrill seekers should always select an outer horse.

The center of mass is the point that provides the average location for the total mass of a system. The word "system" can apply to just one object/particle or to many. The center of mass for a system can be calculated by finding the average of the mass of each object and multiplying by its distance from an origin point using the following formula:

$$x_{centerofmass} = \frac{m_1 x_1 + m_2 x_2}{m_1 + m_2}$$

In this case, x is the distance from the point of origin for the center of mass and each respective object, and m is the mass of each object.

To calculate for more than one object, the pattern can be continued by adding additional masses and their respective distances from the origin point.

Simple Machines
A simple machine is a mechanical device that changes the direction or magnitude of a force. There are six basic types of simple machines: lever, wedge, screw, inclined plane, wheel and axle, and pulley.

Here is how each type works and an example:

- A lever helps lift heavy items higher with less force, such as a crowbar lifting a large cast iron lid.

- A wedge helps apply force to a specific area by focusing the pressure, such as an axe splitting a tree.

- An inclined plane, such as a loading dock ramp, helps move heavy items up vertical distances with less force.

- A screw is an inclined plane wrapped around an axis and allows more force to be applied by extending the distance of the plane. For example, a screw being turned into a piece of wood provides greater securing strength than hitting a nail into the wood.

- A wheel and axle allows the use of rotational force around an axis to assist with applying force. For example, a wheelbarrow makes it easier to haul large loads by employing a wheel and axle at the front.

- A pulley is an application of a wheel and axle with the addition of cords or ropes and it helps move objects vertically. For example, pulling a bucket out of a well is easier with a pulley and ropes.

wheel and axle

pulley

wedge

pry bar

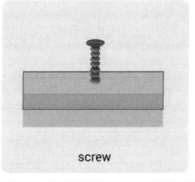
screw

inclined plane

Using a simple machine employs an advantage to the user. This is referred to as the mechanical advantage. It can be calculated by comparing the force input by the user to the simple machine with the force output from the use of the machine (also displayed as a ratio).

$$Mechanical\ Advantage\ =\ \frac{output\ force}{input\ force}$$

$$MA\ =\ \frac{F_{out}}{F_{in}}$$

In the following instance of using a lever, it can be helpful to calculate the torque, or circular force, necessary to move something. This is also employed when using a wrench to loosen a bolt.

$$Torque\ =\ F\ \times\ distance\ of\ lever\ arm\ from\ the\ axis\ of\ rotation\ (called\ the\ moment\ arm)$$

$$T\ =\ F\ \times\ d$$

Electricity and Magnetism

Electrical Nature of Common Materials

Generally, an atom carries no net charge because the positive charges of the protons in the nucleus balance the negative charges of the electrons in the outer shells of the atom. This is considered to be electrically neutral. However, since electrons are the only portion of the atom known to have the freedom to "move," this can cause an object to become electrically charged. This happens either through a gain or a loss of electrons. Electrons have a negative charge, so a gain creates a net negative charge for the object. On the contrary, a loss of electrons creates a positive charge for the object. This charge can also be focused on specific areas of an object, causing a notable interaction between charged objects. For example, if a person rubs a balloon on a carpet, the balloon transfers some of is electrons to the carpet. So, if that person were to hold a balloon near his or her hair, the electrons in the "neutral" hair would make the hair stand on end. This is due to the electrons wanting to fill the deficit of electrons on the balloon. Unless electrically forced into a charged state, most natural objects in nature tend toward reestablishing and maintaining a neutral charge.

When dealing with charges, it is easiest to remember that *like charges repel* each other and *opposite charges attract* each other. Therefore, negatives and positives attract, while two positives or two negatives will repel each other. Similarly, when two charges come near each other, they exert a force on one another. This is described through *Coulomb's Law*:

$$F\ =\ k\frac{q_1 q_2}{r^2}$$

In this equation, *F* is equal to the force exerted by the interaction, *k* is a constant ($k = 8.99 \times 10^9$ N m^2/C^2), q_1 and q_2 are the measure of the two charges, and *r* is the distance between the two charges.

When materials readily transfer electricity or electrons, or can easily accept or lose electrons, they are considered to be good conductors. The transferring of electricity is called *conductivity*. If a material does not readily accept the transfer of electrons or readily loses electrons, it is considered to be an *insulator*. For example, copper wire easily transfers electricity because copper is a good conductor. However, plastic does not transfer electricity because it is not a good conductor. In fact, plastic is an insulator.

Basic Electrical Concepts

In an electrical circuit, the flow from a power source, or the voltage, is "drawn" across the components in the circuit from the positive end to the negative end. This flow of charge creates an electric current (*I*), which is the time (*t*) rate of flow of net charge (*q*). It is measured with the formula:

$$I = \frac{q}{t}$$

Current is measured in amperes (amps). There are two main types of currents:

1. *Direct current* (DC): a unidirectional flow of charges through a circuit

2. *Alternating current* (AC): a circuit with a changing directional flow of charges or magnitude

Every circuit will show a loss in voltage across its conducting material. This loss of voltage is from resistance within the circuit and can be caused by multiple factors, including resistance from wiring and components such as light bulbs and switches. To measure the resistance in a given circuit, Ohm's law is used:

$$Resistance = \frac{Voltage}{current} = R = \frac{V}{I}$$

Resistance (*R*) is measured in Ohms (Ω).

Components in a circuit can be wired *in series* or *in parallel*. If the components are wired in series, a single wire connects each component to the next in line. If the components are wired in parallel, two wires connect each component to the next. The main difference is that the voltage across those in series is directly related from one component to the next. Therefore, if the first component in the series becomes inoperable, no voltage can get to the other components. Conversely, the components in parallel share the voltage across each other and are not dependent on the prior component wired to allow the voltage across the wire.

To calculate the resistance of circuit components wired in series or parallel, the following equations are used:

Resistance in series:

$$R_{total} = R_1 + R_2 + R_3 + \cdots$$

Resistance in parallel:

$$R_{total} = \frac{1}{R_1} + \frac{1}{R_2} + \frac{1}{R_3} + \cdots$$

To make electrons move so that they can carry their charge, a change in voltage must be present. On a small scale, this is demonstrated through the electrons traveling from the light switch to a person's finger. This might happen in a situation where a person runs his or her socks on a carpet, touches a light switch, and receives a small jolt from the electrons that run from the switch to the finger. This minor jolt is due to the deficit of electrons created by rubbing the socks on the carpet, and then the electrons going into the ground. The difference in charge between the switch and the finger caused the electrons to move.

If this situation were to be created on a larger and more sustained scale, the factors would need to be more systematic, predictable, and harnessed. This could be achieved through batteries/cells and generators. Batteries or cells have a chemical reaction that occurs inside, causing energy to be released and charges to be able to move freely. Batteries generally have nodes (one positive and one negative), where items can be hooked up to complete a circuit and allow the charge to travel freely through the item. Generators convert mechanical energy into electric energy using power and movement.

Basic Properties of Magnetic Fields and Forces
Consider two straight rods that are made from magnetic material. They will naturally have a negative end (pole) and a positive end (pole). These charged poles react just like any charged item: opposite charges attract and like charges repel. They will attract each other when arranged positive pole to negative pole. However, if one rod is turned around, the two rods will now repel each other due to the alignment of negative to negative and positive to positive. These types of forces can also be created and amplified by using an electric current. For example, sending an electric current through a stretch of wire creates an electromagnetic force around the wire from the charge of the current. This force exists as long as the flow of electricity is sustained. This magnetic force can also attract and repel other items with magnetic properties. Depending on the strength of the current in the wire, a greater or smaller magnetic force can be generated around the wire. As soon as the current is stopped, the magnetic force also stops.

Optics and Waves
Electromagnetic Spectrum
The movement of light is described like the movement of waves. Light travels with a wave front, has an amplitude (height from the neutral), a cycle or wavelength, a period, and energy. Light travels at approximately 3.00×10^8 m/s and is faster than anything created by humans thus far.

Light is commonly referred to by its measured wavelengths, or the distance between two successive crests or troughs in a wave. Types of light with the longest wavelengths include radio, TV, and micro, and infrared waves. The next set of wavelengths are detectable by the human eye and create the *visible spectrum*. The visible spectrum has wavelengths of 10^{-7} m, and the colors seen are red, orange, yellow, green, blue, indigo, and violet. Beyond the visible spectrum are shorter wavelengths (also called the *electromagnetic spectrum*) containing ultraviolet light, X-rays, and gamma rays. The wavelengths outside of the visible light range can be harmful to humans if they are directly exposed or are exposed for long periods of time. For example, the light from the Sun has a small percentage of ultraviolet (UV) light, which is mostly absorbed by the UV layer of the Earth's atmosphere. When this layer does not filter out the UV rays, the exposure to the wavelengths can be harmful to humans' skin. When there is an extra layer of pollutants, and the light from the sun is trapped by repeated reflection to the Earth (so that it is unable to bounce back into space), it creates another harmful condition for the Earth called the *greenhouse effect*. This is an overexposure to the Sun's light and contributes to *global warming* by increasing the temperature on Earth.

Basic Characteristics and Types of Waves
A *mechanical wave* is a type of wave that passes through a medium (solid, liquid, or gas). There are two basic types of mechanical waves: longitudinal and transverse.

A *longitudinal wave* has motion that is parallel to the direction of the wave's travel. This can best be visualized by compressing one side of a tethered spring and then releasing that end. The movement travels in a bunching/un-bunching motion across the length of the spring and back.

A *transverse wave* has motion that is perpendicular to the direction of the wave's travel. The particles on a transverse wave do not move across the length of the wave; instead, they oscillate up and down, creating peaks and troughs.

A wave with a combination of both longitudinal and transverse motion can be seen through the motion of a wave on the ocean—with peaks and troughs, and particles oscillating up and down.

Mechanical waves can carry energy, sound, and light, but they need a medium through which transport can occur. An electromagnetic wave can transmit energy without a medium, or in a vacuum.

A more recent addition in the study of waves is the *gravitational wave*. Its existence has been proven and verified, yet the details surrounding its capabilities are still somewhat under inquiry. Gravitational waves are purported to be ripples that propagate as waves outward from their source and travel in the curvature of space/time. They are thought to carry energy in a form of radiant energy called *gravitational radiation*.

Basic Wave Phenomena

When a wave crosses a boundary or travels from one medium to another, certain things occur. If the wave can travel through one medium into another medium, it experiences *refraction*. This is the bending of the wave from one medium to another due to a change in density of the mediums, and thus, the speed of the wave changes. For example, when a pencil is sitting in half of a glass of water, a side view of the glass makes the pencil appear to be bent at the water level. What the viewer is seeing is the refraction of light waves traveling from the air into the water. Since the wave speed is slowed in water, the change makes the pencil appear bent.

When a wave hits a medium that it cannot penetrate, it is bounced back in an action called *reflection*. For example, when light waves hit a mirror, they are reflected, or bounced, off the mirror. This can cause it to seem like there is more light in the room, since there is a "doubling back" of the initial wave. This same phenomenon also causes people to be able to see their reflection in a mirror.

When a wave travels through a slit or around an obstacle, it is known as *diffraction*. A light wave will bend around an obstacle or through a slit and cause what is called a *diffraction pattern*. When the waves bend around an obstacle, it causes the addition of waves and the spreading of light on the other side of the opening.

Dispersion is used to describe the splitting of a single wave by refracting its components into separate parts. For example, if a wave of white light is sent through a dispersion prism, the light appears as its separate rainbow-colored components, due to each colored wavelength being refracted in the prism.

When wavelengths hit boundaries, different things occur. Objects will absorb certain wavelengths of light and reflect others, depending on the boundaries. This becomes important when an object appears to be a certain color. The color of an object is not actually within that object, but rather, in the wavelengths being transmitted by that object. For example, if a table appears to be red, that means the table is absorbing all other wavelengths of visible light except those of the red wavelength. The table is reflecting, or transmitting, the wavelengths associated with red back to the human eye, and so it appears red.

Interference describes when an object affects the path of a wave, or another wave interacts with a wave. Waves interacting with each other can result in either *constructive interference* or *destructive interference*, based on their positions. With constructive interference, the waves are in sync with each other and combine to reinforce each other. In the case of deconstructive interference, the waves are out of sync and reduce the effect of each other to some degree. In *scattering*, the boundary can change the direction or energy of a wave, thus altering the entire wave. *Polarization* changes the oscillations of a wave and can alter its appearance in light waves. For example, polarized sunglasses remove the "glare" from sunlight by altering the oscillation pattern observed by the wearer.

When a wave hits a boundary and is completely reflected, or if it cannot escape from one medium to another, it is called *total internal reflection*. This effect can be seen in the diamonds with a brilliant cut. The angle cut on the sides of the diamond causes the light hitting the diamond to be completely reflected back inside the gem, making it appear brighter and more colorful than a diamond with different angles cut into its surface.

The *Doppler effect* applies to situations with both light and sound waves. The premise of the Doppler effect is that, based upon the relative position or movement of a source and an observer, waves can seem shorter or longer than they actually are. When the Doppler effect is noted with sound, it warps the noise being heard by the observer. This makes the pitch or frequency seem shorter or higher as the source is approaching, and then longer or lower as the source is getting farther away. The frequency/pitch of the source never actually changes, but the sound in respect to the observer makes it seem like the sound has changed. This can be observed when a siren passes by an observer on the road. The siren sounds much higher in pitch as it approaches the observer and then lower after it passes and is getting farther away.

The Doppler effect also applies to situations involving light waves. An observer in space would see light approaching as being shorter wavelengths than the light actually is, causing it to look blue. When the light wave gets farther away, the light would appear red because of the apparent elongation of the wavelength. This is called the *red-blue shift*.

Basic Optics

When reflecting light, a mirror can be used to observe a virtual (not real) image. A *plane mirror* is a piece of glass with a coating in the background to create a reflective surface. An image is what the human eye sees when light is reflected off the mirror in an unmagnified manner. If a *curved mirror* is used for reflection, the image seen will not be a true reflection. Instead, the image will either be enlarged or miniaturized compared to its actual size. Curved mirrors can also make the object appear closer or farther away than the actual distance the object is from the mirror.

Lenses can be used to refract or bend light to form images. Examples of lenses are the human eye, microscopes, and telescopes. The human eye interprets the refraction of light into images that humans understand to be actual size. *Microscopes* allow objects that are too small for the unaided human eye to be enlarged enough to be seen. *Telescopes* allow objects to be viewed that are too far away to be seen with the unaided eye. *Prisms* are pieces of glass that can have a wavelength of light enter one side and appear to be divided into its component wavelengths on the other side. This is due to the ability of the prism to slow certain wavelengths more than others.

Sound

Sound travels in waves and is the movement of vibrations through a medium. It can travel through air (gas), land, water, etc. For example, the noise a human hears in the air is the vibration of the waves as they reach the ear. The human brain translates the different frequencies (pitches) and intensities of the vibrations to determine what created the noise.

A tuning fork has a predetermined frequency because of the length and thickness of its tines. When struck, it allows vibrations between the two tines to move the air at a specific rate. This creates a specific tone, or note, for that size of tuning fork. The number of vibrations over time is also steady for that tuning fork and can be matched with a frequency. All pitches heard by the human ear are categorized by using frequency and are measured in Hertz (cycles per second).

The level of sound in the air is measured with sound level meters on a decibel (dB) scale. These meters respond to changes in air pressure caused by sound waves and measure sound intensity. One decibel is 1/10th of a *bel*, named after Alexander Graham Bell, the inventor of the telephone. The decibel scale is logarithmic, so it is measured in factors of 10. This means, for example, that a 10 dB increase on a sound meter equates to a 10-fold increase in sound intensity.

Life Science

Function of Cells

Structure and Function of Cell Membranes
All cells are surrounded by a cell membrane that is formed from two layers of phospholipids. *Phospholipids* are two fatty acid chains connected to a glycerol molecule with a phosphate group. The membrane is amphiphilic because the fatty acid chains are hydrophobic and the phosphate group is hydrophilic. This creates a unique environment that protects the cell's inner contents while still allowing material to pass through the membrane. Because the outside of a cell, known as the *extracellular space*, and the inside of a cell, the *intercellular space*, are aqueous, the lipid bilayer forms with the two phospholipid heads facing the outside and the inside of the cell. This allows the phospholipids to interact with water; the fatty acid tails face the middle, so they can interact with each other and avoid water.

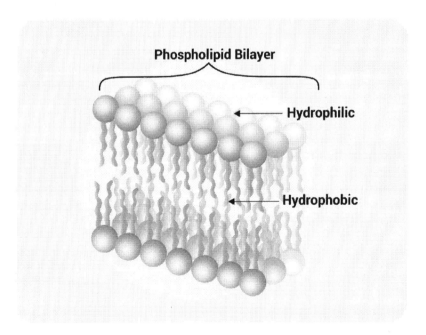

Molecules can pass through the cell membrane by either active or passive transport. *Active transport* requires chemical energy to move molecules in or out of the cell from areas of low concentration to areas of high concentration, or in instances where the molecules would not be able to pass through by themselves, such as with large non-lipid soluble molecules. Ions, amino acids, and complex sugars use active transport mechanisms. Molecules that are soluble in lipids, water, and oxygen use *passive transport* to move in and out of the cell, which means that cellular energy is not required for their movement. Examples of passive transport include diffusion, facilitated diffusion, and osmosis. *Diffusion* is the net movement of particles from an area of high concentration to lower concentration. *Facilitated diffusion* is the movement of molecules through cell membranes with the use of special transport proteins. Finally, *osmosis* is the movement of water molecules across partially permeable membranes.

<u>Structure and Function of Animal and Plant Cell Organelles</u>
Animal and plant cells contain many of the same or similar *organelles*, which are membrane enclosed structures that each have a specific function; however, there are a few organelles that are unique to either one or the other general cell type. The following cell organelles are found in both animal and plant cells, unless otherwise noted in their description:

- *Nucleus*: The nucleus consists of three parts: the nuclear envelope, the nucleolus, and chromatin. The *nuclear envelope* is the double membrane that surrounds the nucleus and separates its contents from the rest of the cell. The *nucleolus* produces ribosomes. *Chromatin* consists of DNA and protein, which form chromosomes that contain genetic information. Most cells have only one nucleus; however, some cells, such as skeletal muscle cells, have multiple nuclei.

- *Endoplasmic reticulum (ER)*: The ER is a network of membranous sacs and tubes that is responsible for membrane synthesis. It is also responsible for packaging and transporting proteins into vesicles that can move out of the cell. It folds and transports other proteins to the Golgi apparatus. It contains both smooth and rough regions; the rough regions have ribosomes attached, which are the sites of protein synthesis.

- *Flagellum*: Flagellum are found only in animal cells. They are made up of a cluster of microtubules projected out of the plasma membrane, and they aid in cell mobility.

- *Centrosome*: The centrosome is the area of the cell where *microtubules*, which are filaments that are responsible for movement in the cell, begin to be formed. Each centrosome contains two centrioles. Each cell contains one centrosome.

- *Cytoskeleton*: The cytoskeleton in animal cells is made up of microfilaments, intermediate filaments, and microtubules. In plant cells, the cytoskeleton is made up of only microfilaments and microtubules. These structures reinforce the cell's shape and aid in cell movement.

- *Microvilli*: Microvilli are found only in animal cells. They are protrusions in the cell membrane that increase the cell's surface area. They have a variety of functions, including absorption, secretion, and cellular adhesion. They are found on the apical surface of epithelial cells, such as in the small intestine. They are also located on the plasma surface of a female's eggs to help anchor sperm that are attempting fertilization.

- *Peroxisome*: A peroxisome contains enzymes that are involved in many of the cell's metabolic functions, one of the most important being the breakdown of very long chain fatty acids. Peroxisomes produces hydrogen peroxide as a byproduct of these processes and then converts the hydrogen peroxide to water. There are many peroxisomes in each cell.

- *Mitochondrion*: The mitochondrion is often called the powerhouse of the cell and is one of the most important structures for maintaining regular cell function. It is where aerobic cellular respiration occurs and where most of the cell's adenosine triphosphate (ATP) is generated. The number of mitochondria in a cell varies greatly from organism to organism, and from cell to cell. In human cells, the number of mitochondria can vary from zero in a red blood cell, to 2000 in a liver cell.

- *Lysosome*: Lysosomes are exclusively found in animal cells. They are responsible for digestion and can hydrolyze macromolecules. There are many lysosomes in each cell.

- *Golgi apparatus*: The Golgi apparatus is responsible for the composition, modification, organization, and secretion of cell products. Because of its large size, it was actually one of the first organelles to be studied in detail. There are many Golgi apparati in each cell.

- *Ribosomes*: Ribosomes are found either free in the cytosol, bound to the rough ER, or bound to the nuclear envelope. They manufacture proteins within the cell.

- *Plasmodesmata*: The plasmodesmata are found only in plant cells. They are cytoplasmic channels, or tunnels, that go through the cell wall and connect the cytoplasm of adjacent cells.

- *Chloroplast*: Chloroplasts are found only in plant cells. They are responsible for *photosynthesis*, which is the process of converting sunlight to chemical energy that can be stored and used later to drive cellular activities.

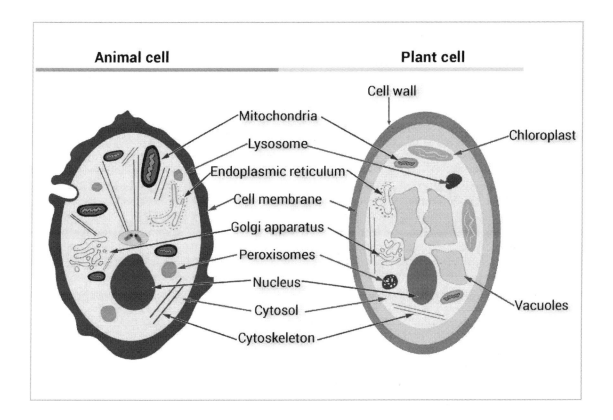

- *Central vacuole*: A central vacuole is found only in plant cells. It is responsible for storing material and waste. This is the only vacuole found in a plant cell.

- *Plasma membrane*: The plasma membrane is a phospholipid bilayer that encloses the cell.

- *Cell wall*: Cell walls are only present in plant cells. The cell wall is made up of strong fibrous substances including cellulose and other polysaccharides, and protein. It is a layer outside of the plasma membrane, which protects the cell from mechanical damage and helps maintain the cell's shape.

Levels of Organization

There are about two hundred different types of cells in the human body. Cells group together to form *biological tissues*, and tissues combine to form *organs*, such as the heart and kidneys. Organs that work together to perform vital functions of the human body form *organ systems*. There are eleven organ systems in the human body: skeletal, muscular, urinary, nervous, digestive, endocrine, reproductive, respiratory, cardiovascular, integumentary, and lymphatic. Although each system has its own unique function, they all rely on each other, either directly or indirectly, to operate properly.

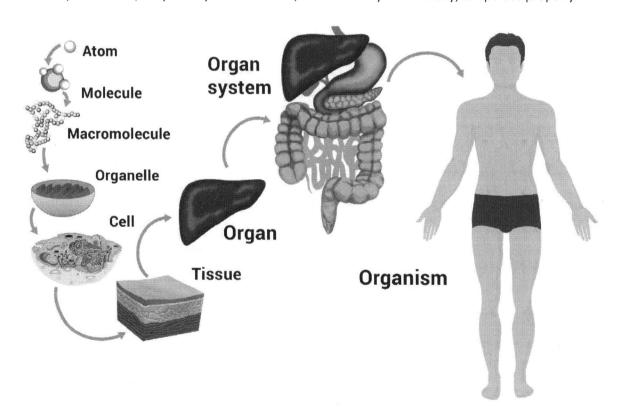

Major Features of Common Animal Cell Types

The most common animal cell types are blood, muscle, nerve, epithelial, and gamete cells. The three main blood cells are *red blood cells (RBCs), white blood cells (WBCs),* and *platelets*. RBCs transport oxygen and carbon dioxide through the body. They do not have a nucleus and they live for about 120 days in the blood. WBCs defend the body against diseases. They do have a nucleus and live for only three to four days in the human body. Platelets help with the formation of blood clots following an injury. They do not have a nucleus and live for about eight days after formation. *Muscle cells* are long, tubular cells that form muscles, which are responsible for movement in the body. On average, they live for about fifteen years, but this number is highly dependent on the individual body. There are three main types of muscle tissue: skeletal, cardiac, and smooth. *Skeletal muscle cells* have multiple nuclei and are the only voluntary muscle cell, which means that the brain consciously controls the movement of skeletal muscle. *Cardiac muscle cells* are only found in the heart; they have a single nucleus and are involuntary. *Smooth muscle cells* make up the walls of the blood vessels and organs. They have a single nucleus and are involuntary. *Nerve cells* conduct electrical impulses that help send information and instructions from the brain to the rest of the body. They contain a single nucleus and have a specialized membrane that allows for this electrical signaling between cells. *Epithelial* cells cover exposed surfaces, and line internal cavities and passageways. *Gametes* are specialized cells that are responsible for reproduction. In the human body, the gametes are the egg and the sperm.

Prokaryotes and Eukaryotes

There are two distinct types of cells that make up most living organisms: *prokaryotic* and *eukaryotic*. Both types of cells are enclosed by plasma membranes with cytosol on the inside. They both contain *ribosomes* and DNA. One major difference between these types of cells is that in eukaryotic cells, the cell's DNA is enclosed in a membrane-bound nucleus, whereas in prokaryotic cells, the cell's DNA is in a region—called the *nucleoid*—that is not enclosed by a membrane. Another major difference is that eukaryotic cells contain organelles, while prokaryotic cells do not have organelles.

Prokaryotic cells include *bacteria* and archaea. They do not have a nucleus or any membrane-bound organelles, are unicellular organisms, and are generally very small in size. Eukaryotic cells include animal, plant, fungus, and protist cells. *Fungi* are unicellular microorganisms such as yeasts, molds, and mushrooms. Their distinguishing characteristic is the chitin that is in their cell walls. *Protists* are organisms that are not classified as animals, plants, or fungi; they are unicellular; and they do not form tissues.

Key Aspects of Cell Reproduction and Division

<u>Cell Cycle</u>

The *cell cycle* is the process by which a cell divides and duplicates itself. There are two processes by which a cell can divide itself: mitosis and meiosis. In *mitosis*, the daughter cells that are produced from parental cell division are identical to each other and the parent. *Meiosis* is a unique process that involves two stages of cell division and produces *haploid cells*, which are cells containing only one set of chromosomes, from *diploid parent cells*, which are cells containing two sets of chromosomes.

The Cell Cycle

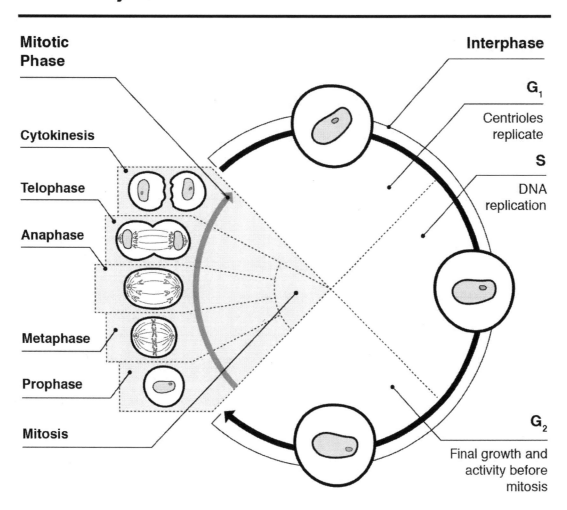

Mitotic Phase

Cytokinesis

Telophase

Anaphase

Metaphase

Prophase

Mitosis

Interphase

G_1

Centrioles replicate

S

DNA replication

G_2

Final growth and activity before mitosis

Mitosis
Mitosis can be broken down into five stages: prophase, prometaphase, metaphase, anaphase, and telophase.

- *Prophase*: During this phase, the mitotic spindles begin to form from centrosomes and microtubules. As the microtubules lengthen, the centrosomes move farther away from each other. The nucleolus disappears and the chromatin fibers begin to coil up and form chromosomes. Two sister chromatids, which are two copies of one chromosome, are joined together.

- *Prometaphase*: The nuclear envelope begins to break down and the microtubules enter the nuclear area. Each pair of chromatin fibers develops a *kinetochore*, which is a specialized protein structure in the middle of the adjoined fibers. The chromosomes are further condensed.

- *Metaphase*: In this stage, the microtubules are stretched across the cell and the centrosomes are at opposite ends of the cell. The chromosomes align at the *metaphase plate*, which is a plane that is exactly between the two centrosomes. The kinetochore of each chromosome is attached to the kinetochore of the microtubules that are stretching from each centrosome to the metaphase plate.

- *Anaphase*: The sister chromatids break apart, forming full-fledged chromosomes. The two daughter chromosomes move to opposite ends of the cell. The microtubules shorten toward opposite ends of the cell as well, and the cell elongates.

- *Telophase*: Two nuclei form at each end of the cell and nuclear envelopes begin to form around each nucleus. The nucleoli reappear and the chromosomes become less condensed. The microtubules are broken down by the cell and mitosis is complete.

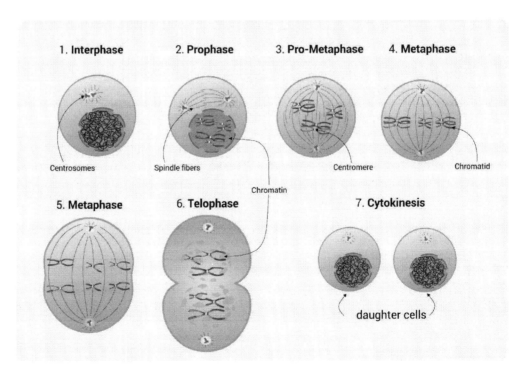

<u>Meiosis</u>
Meiosis is a type of cell division in which the daughter cells have half as many sets of chromosomes as the parent cell. In addition, one parent cell produces four daughter cells. Meiosis has the same phases as mitosis, except that they occur twice—once in meiosis I and once in meiosis II. The diploid parent has two sets of chromosomes, set A and set B. During meiosis I, each chromosome set duplicates, producing a second set of A chromosomes and a second set of B chromosomes, and the cell splits into two. Each cell contains two sets of chromosomes. Next, during meiosis II, the two intermediate daughter cells divide again, producing four total haploid cells that each contain one set of chromosomes. Two of the haploid cells each contain one chromosome of set A and the other two cells each contain one chromosome of set B.

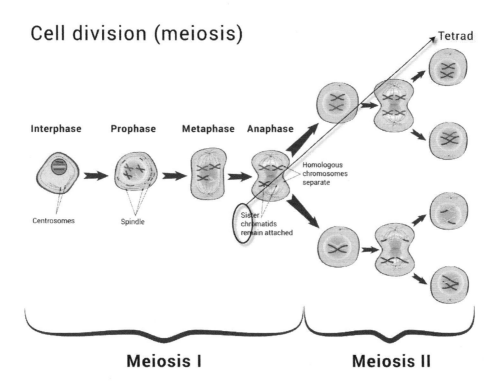

<u>Cytokinesis</u>
Cytokinesis is the division of cytoplasm that occurs immediately following the division of genetic material during cellular reproduction. The process of mitosis or meiosis, followed by cytokinesis, makes up the complete cell cycle.

Basic Biochemistry of Life

<u>Cellular Respiration</u>
Cellular respiration is a set of metabolic processes that converts energy from nutrients into ATP, which is the molecule of useable energy for the cell. Respiration can either occur aerobically, using oxygen, or anaerobically, without oxygen. While prokaryotic cells carry out respiration in the cytosol, most of the aerobic respiration in eukaryotic cells occurs in the mitochondria. Glycolysis and ATP-PC (phosphocreatine system) take place in the cytosol.

Anaerobic Respiration

Some organisms do not live in oxygen-rich environments and must find alternate methods of respiration. *Anaerobic respiration* occurs in certain prokaryotic organisms, and while it does occur in eukaryotic organisms, it happens in them much less frequently. The organisms utilize an electron transport chain similar to that of the aerobic respiration pathway; the terminal acceptor molecule, however, is an electronegative substance that is not an oxygen molecule. Some bacteria, for example, use the sulfate ion (SO_4^{2-}) as the final electron accepting molecule and the resulting byproduct is hydrogen sulfide (H_2S), instead of water.

Aerobic Respiration

There are two main steps in *aerobic cellular respiration*: the *citric acid cycle*, also known as the *Krebs cycle*, and *oxidative phosphorylation*. A process called *glycolysis* converts glucose molecules into pyruvate molecules and those pyruvate molecules then enter the citric acid cycle. The pyruvate molecules are broken down to produce ATP, as well as NADH and $FADH_2$—molecules that are used energetically to drive the next step of oxidative phosphorylation. During this phase of aerobic respiration, an electron transport chain pumps electrons and protons across the inner mitochondrial matrix. The electrons are accepted by an oxygen molecule, and water is produced. This process then fuels *chemiosmosis*, which helps convert ADP molecules to ATP. The total number of ATP molecules generated through aerobic respiration can be as many as thirty-eight, if none are lost during the process. Aerobic respiration is up to fifteen times more efficient than anaerobic respiration.

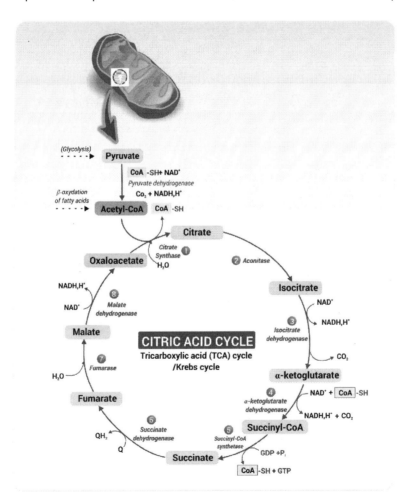

Photosynthesis

Photosynthesis is the process of converting light energy into chemical energy, which is then stored in sugar and other organic molecules. It can be divided into two stages called the *light reactions* and the *Calvin cycle*. The photosynthetic process takes place in the chloroplast in plants. Inside the chloroplast, there are membranous sacs called *thylakoids*. *Chlorophyll* is a green pigment that lives in the thylakoid membranes, absorbs photons from light, and starts an electron transport chain in order to produce energy in the form of ATP and NADPH. The ATP and NADPH produced from the light reactions are used as energy to form organic molecules in the Calvin cycle.

The Calvin cycle takes place in the *stroma*, or inner space, of the chloroplasts. The process consumes nine ATP molecules and six NADPH molecules for every one molecule of glyceraldehyde 3-phosphate (G3P) that it produces. The G3P that is produced can be used as the starting material to build larger organic compounds, such as glucose. The complex series of reactions that takes place in photosynthesis can be simplified into the following equation: $6CO_2 + 12 H_2O + \text{Light Energy} \rightarrow C_6H_{12}O_6 + 6O_2 + 6H_2O$.

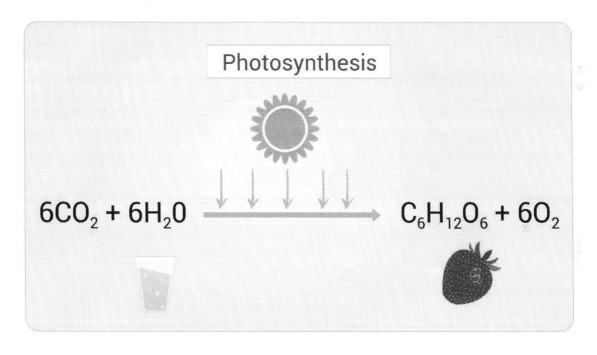

Basically, carbon dioxide and water mix with light energy inside the chloroplast to produce organic molecules, oxygen, and water. It is interesting to note that water is on both sides of the equation. Twelve water molecules are consumed during this process and six water molecules are newly formed as byproducts. Although the Calvin cycle itself is not dependent on light energy, both steps of photosynthesis usually occur during daylight because the Calvin cycle is dependent upon the ATP and NADPH that is produced by the light reactions.

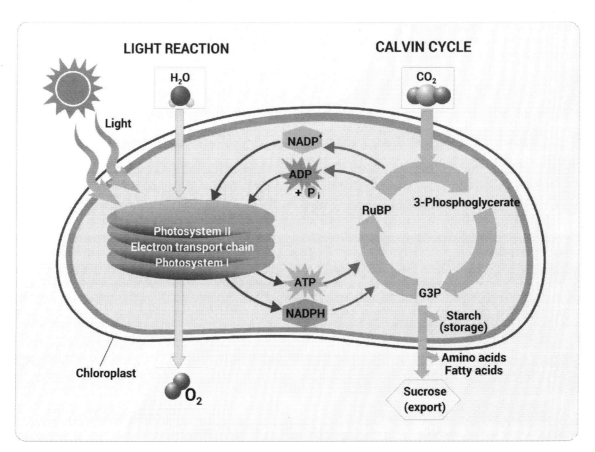

Biological Molecules
Repeating units of monomers (small molecules that bond with identical small molecules) that are linked together are called *polymers*. The most important polymers found in all living things can be divided into five categories: nucleic acids (such as DNA), carbohydrates, proteins, lipids, and enzymes. Carbon (C), hydrogen (H), oxygen (O), nitrogen (N), sulfur (S), and phosphorus (P) are the major elements of most biological molecules. Carbon is a common backbone of large molecules because of its ability to bond to four different atoms.

DNA and RNA
Nucleotides consist of a five-carbon sugar, a nitrogen-containing base, and one or more phosphate groups. *Deoxyribonucleic acid (DNA)* is made up of two strands of nucleotides coiled together in a double-helix structure. It plays a major role in enabling living organisms to pass their genetic information and complex components on to subsequent generations. There are four nitrogenous bases that make up DNA: adenine, thymine, guanine, and cytosine. Adenine always pairs with thymine, and guanine always pairs with cytosine. *Ribonucleic acid (RNA)* is often made up of only one

strand of nucleotides folded in on itself. Like DNA, RNA has four nitrogenous bases; however, in RNA, thymine is replaced by uracil.

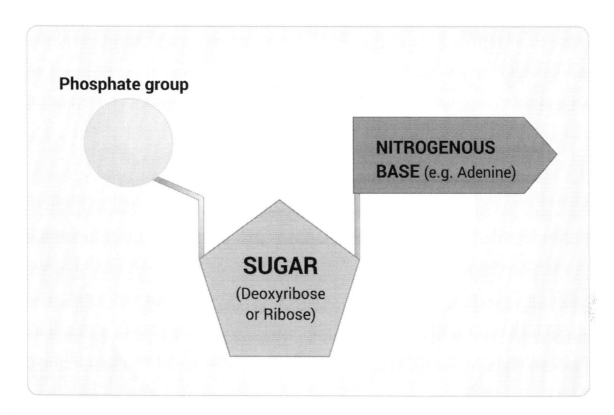

Carbohydrates

Carbohydrates consist of sugars and polymers of sugars, such as starches, which make up the cell walls of plants. The simplest sugar is called a *monosaccharide* and has the molecular formula of CH_2O, or a multiple of that formula. Monosaccharides are important molecules for cellular respiration. Their carbon skeleton can also be used to rebuild new small molecules. *Polysaccharides* are made up of a few hundred to a few thousand monosaccharides linked together.

Proteins

Proteins are essential for almost all functions in living beings. All proteins are made from a set of twenty *amino acids* that are linked in *unbranched polymers*. The amino acids are linked by *peptide bonds*, and polymers of amino acids are called *polypeptides*. These polypeptides, either individually or in linked combination with each other, fold up and form coils of biologically functional molecules.

There are four levels of protein structure: primary, secondary, tertiary, and quaternary. The *primary structure* is the sequence of amino acids, similar to the letters in a long word. The *secondary structure* comprises the folds and coils that are formed by hydrogen bonding between the slightly charged atoms of the polypeptide backbone. *Tertiary structure* is the overall shape of the molecule that results from the interactions between the side chains that are linked to the polypeptide backbone. *Quaternary structure* is the overall protein structure that occurs when a protein is made up of two or more polypeptide chains.

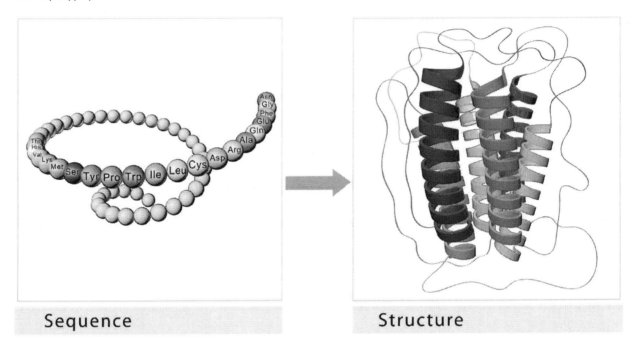

Sequence   Structure

Lipids

Lipids are a class of biological molecules that are *hydrophobic*, which means that they do not mix well with water. They are mostly made up of large chains of carbon and hydrogen atoms, termed *hydrocarbon chains*. The three most important types of lipids are fats, phospholipids, and steroids.

Fats are made up of two types of smaller molecules: three fatty acids and one glycerol molecule. Saturated fats do not have double bonds between the carbons in the fatty acid chain, such as glycerol, pictured below. They are fairly straight molecules and can pack together closely, so they form solids at room temperature. Unsaturated fats have one or more double bonds between carbons in the fatty acid chain. Since they cannot pack together as tightly as saturated fats, they take up more space and are called oils. They remain liquid at room temperature.

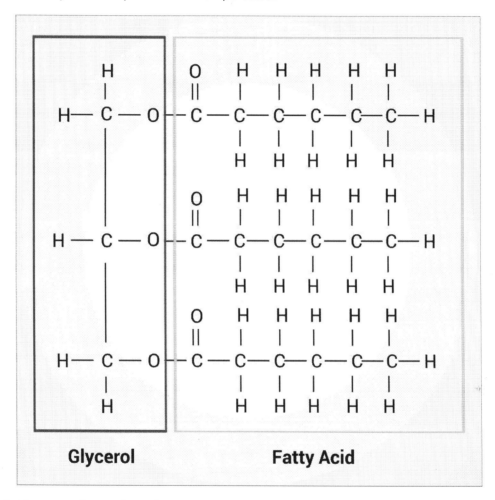

Glycerol   **Fatty Acid**

Phospholipids are made up of two fatty acid molecules linked to one glycerol molecule. When phospholipids are mixed with water, they inherently create double-layered structures, called *bilayers*, which shield their hydrophobic regions from the water molecules.

Steroids are lipids that consist of four fused carbon rings. They can mix in between the phospholipid bilayer cell membrane and help maintain its structure, as well as aid in cell signaling.

Enzymes
Enzymes are biological molecules that accelerate the rate of chemical reactions by lowering the activation energy needed to make the reaction proceed. Although most enzymes can be classified as proteins, some are ribonucleic acid (RNA) molecules. Enzymes function by interacting with a specific substrate in order to create a different molecule, or product. Most reactions in cells need enzymes to make them occur at rates fast enough to sustain life.

Basic Genetics

Structure and Function of DNA and RNA
DNA and RNA are made up of *nucleotides*, which are formed from a five-carbon sugar, a nitrogenous base, and one or more phosphate group. While DNA is made up of the sugar deoxyribose, RNA is made up of the sugar ribose. Deoxyribose has one fewer oxygen atom than ribose. DNA and RNA each comprise four nitrogenous bases, three of which they have in common: adenine, guanine, and cytosine. Thymine is found only in DNA and uracil is found only in RNA. Each base has a specific pairing formed by hydrogen bonds, and is known as a *base pair*. Adenine interacts with thymine or uracil, and guanine interacts only with cytosine. While RNA is found in a single strand, DNA is a double-stranded molecule that coils up to form a *double helix* structure.

The specific pairing of the nitrogenous bases allows for the hereditary information stored in DNA to be passed down accurately from parent cells to daughter cells. When chromosomes are *replicated* during cell division, the double-helix DNA is first uncoiled, each strand is replicated, and then two new identical DNA molecules are generated. DNA can also be used as a template for generating proteins. A *single-stranded* RNA is generated from the DNA during a process called *transcription*; proteins are then generated from this RNA in a process called *translation*.

Chromosomes, Genes, Alleles
Chromosomes are found inside the nucleus of cells and contain the hereditary information of the cell in the form of *genes*. Each gene has a specific sequence of DNA that eventually encodes proteins and results in inherited traits. *Alleles* are variations of a specific gene that occur at the same location on the chromosome. For example, blue and brown are two different alleles of the gene that encodes for eye color.

Dominant and Recessive Traits
In genetics, *dominant alleles* are mostly noted in capital letters (A) and *recessive alleles* are mostly noted in lower case letters (a). There are three possible combinations of alleles among dominant and recessive alleles: AA, Aa, and aa. Dominant traits are phenotypes that appear when at least one dominant allele is present in the gene. Dominant alleles are considered to have stronger phenotypes and, when mixed with recessive alleles, will mask the recessive trait. The recessive trait would only appear as the phenotype when the allele combination is "aa" because a dominant allele is not present to mask it.

Mendelian Inheritance
A monk named Gregor Mendel is referred to as the father of genetics. He was responsible for coming up with one of the first models of inheritance in the 1860s. His model included two laws to determine which traits are inherited. These laws still apply today, even after genetics has been studied much more in depth.

- *The Law of Segregation*: Each characteristic has two versions that can be inherited. When two parent cells form daughter cells, the two alleles of the gene segregate and each daughter cell can inherit only one of the alleles from each parent.

- *The Law of Independent Assortment*: The alleles for different traits are inherited independent of one another. In other words, the biological selection of one allele by a daughter cell is not linked to the biological selection of an allele for a different trait by the same daughter cell. The genotype that is inherited is the alleles that are encoded on the gene, and the phenotype is the outward appearance of the physical trait for that gene. For example, "A" is the dominant allele for brown eyes and "a" is the recessive allele for blue eyes; the phenotype of brown eyes would occur for two different genotypes: both "AA" and "Aa."

Punnett Squares

For simple genetic combinations, a *Punnett square* can be used to assess the phenotypic ratios of subsequent generations. In a 2 x 2 cell square, one parent's alleles are set up in columns and the other parent's alleles are set up in rows. The resulting allele combinations are shown in the four internal cells.

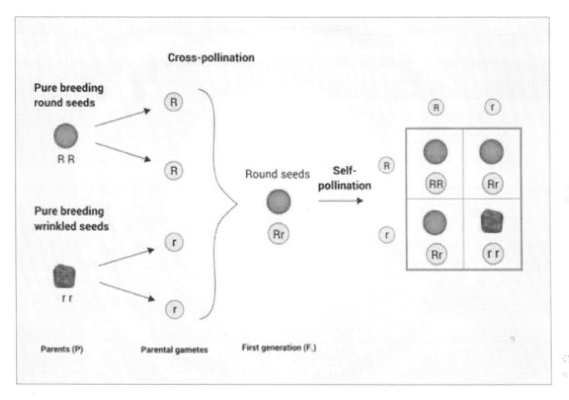

Pedigree

For existing populations where genetic crosses cannot be controlled, phenotype information can be collected over several generations and a *pedigree analysis* can be done to investigate the dominant and recessive characteristics of specific traits. There are several rules to follow when determining the pedigree of a trait. For dominant alleles:

- Affected individuals have at least one affected parent;
- The phenotype appears in every generation; and
- If both parents are unaffected, their offspring will always be unaffected.

For recessive alleles:

- Unaffected parents can have affected offspring; and
- Affected offspring are male and female.

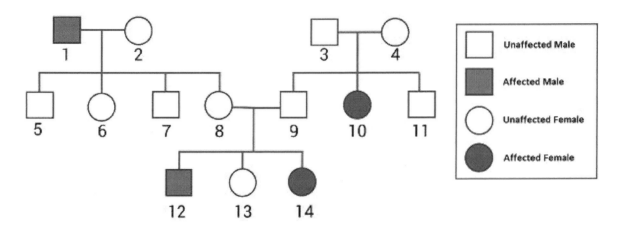

Mutations, Chromosomal Abnormalities, and Common Genetic Disorders

Mutations

Genetic *mutations* occur when there is a permanent alteration in the DNA sequence that codes for a specific gene. They can be small, affecting only one base pair, or large, affecting many genes on a chromosome. Mutations are classified as either hereditary, which means they were also present in the parent gene, or acquired, which means that they occurred after the genes were passed down from the parents. Although mutations are not common, they are an important aspect of genetics and variation in the general population.

Chromosomal Abnormalities and Common Genetic Disorders

Structural chromosomal abnormalities are mutations that affect a large chromosomal segment of more than one gene. This often occurs due to an error in cell division. Acute myelogenous leukemia is caused by a *translocation error*, which is when a segment of one chromosome is moved to another chromosome.

There can also be an abnormal number of chromosomes, which is referred to as *aneuploidy*. Down syndrome is an example of an aneuploidy in which there are three copies of chromosome 21 instead of two copies. Turner syndrome is another example of aneuploidy, in which a female is completely or partially missing an X chromosome. Without the second X chromosome, these females do not develop all of the typical female physical characteristics and are unable to bear children.

Theory and Key Mechanisms of Evolution

Mechanisms of Evolution

Evolution is the concept that there is one common ancestor for all living organisms, and, over time, genetic variation and mutations cause the development of different species. Charles Darwin came up with a scientific model of evolution based on the idea that individuals within a population can have longer lives (better survival) and higher reproduction rates based on certain specific traits that they have inherited, called *natural selection*. The variation of a trait that enhances survival and reproduction in the environment is the one that gets passed on. The survival and inheritance of these traits through many subsequent generations causes a change in the overall population. The traits that are more advantageous for survival and reproduction become more common in subsequent generations and increase the diversity of the population. For example, when there was a drought in the Galapagos Islands, the finches with large beaks became more populous because they were able to survive on the larger, rougher seeds that were remaining.

Speciation and Isolation Methods

Speciation is the method by which one species splits into two or more species due to either geographic separation, called allopatric speciation, or a reduction in gene flow between varying members of the population, called sympatric speciation. In *allopatric speciation*, one population is divided into two subpopulations. For example, if a drought occurs and a large lake becomes divided into two smaller lakes, each lake is left with its own population that cannot intermingle with the population of the other lake. When the genes of these two subpopulations are no longer mixing with each other, new mutations can arise and natural selection can take place.

In *sympatric speciation*, gene flow in the population is reduced by polyploidy, sexual selection, and habitat differentiation. *Polyploidy* is more common in plants than animals and results when cell division during reproduction creates an extra set of chromosomes. In *sexual selection*, organisms of one sex choose their mate of the opposite sex based on certain traits. If there is high selection for two extreme variations of a trait, sympatric speciation may occur. *Habitat differentiation* occurs when a subpopulation exploits a resource that is not used by the parent population. Both allopatric and sympatric speciation can occur quickly or slowly, and may involve just a few gene changes or many gene changes between the new species.

One important distinguishing factor in the formation of two species is their *reproductive isolation*. Species are characterized by their members' ability to breed and produce viable offspring. When speciation occurs and new species are formed, there must have been a biological barrier that prevented the two species from producing viable offspring.

Following speciation, there are two types of *reproductive barriers* that keep the two populations from mating with each other. These are classified as either prezygotic barriers or postzygotic barriers. *Prezygotic barriers* prevent fertilization via habitat isolation, temporal isolation, and behavioral isolation. Through habitat isolation, two species may inhabit the same area but don't often encounter each other. *Temporal isolation* is when species breed at different times of the day, during different seasons, or during different years, so their mating patterns never coincide. *Behavioral isolation* refers to mating rituals that prevent an organism from recognizing a different species as potential mate.

Other prezygotic barriers block fertilization after a mating attempt. *Mechanical isolation* occurs when anatomical differences prevent fertilization. *Gametic isolation* occurs when the gametes of two species are incompatible.

The Fossil Record

Fossils are the preserved remains of animals and organisms from the distant past. They provide evidence of evolution and can elucidate the homology of both living and extinct species. Looking at the *fossil record* over time can help identify how quickly or slowly evolutionary changes occurred, and can also help match those changes to environmental changes that were occurring concurrently.

Comparative Genetics

In *comparative genetics*, different organisms are compared at a genetic level to look for similarities and differences. DNA sequence, genes, gene order, and other structural features are among the features that may be analyzed in order to look for evolutionary relationships and common ancestors between the organisms. Comparative genetics was useful in elucidating the similarities between humans and chimpanzees and linking their evolutionary history.

Homology

Organisms that developed from a common ancestor often have similar characteristics that function differently. This similarity is known as *homology*. For example, humans, cats, whales, and bats all have bones arranged in the same manner from their shoulders to their digits. However, the bones form arms in humans, forelegs in cats, flippers in whales, and wings in bats, and these forelimbs are used for lifting, walking, swimming, and flying, respectively. The similarity of the bone structure shows a common ancestry, but the functional differences are the product of evolution.

Homologous Structures

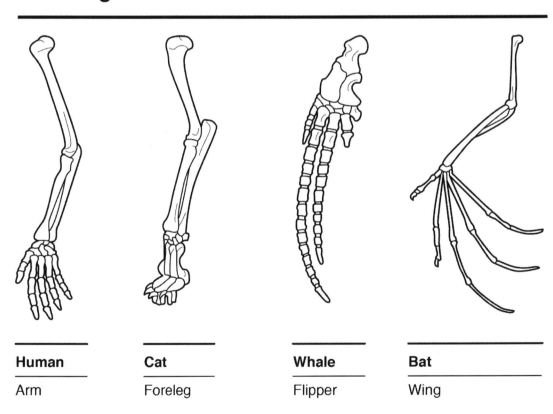

| **Human** | **Cat** | **Whale** | **Bat** |
| Arm | Foreleg | Flipper | Wing |

Hierarchical Classification Schemes

Classification Schemes

Taxonomy is the science behind the biological names of organisms. Biologists often refer to organisms by their Latin scientific names to avoid confusion with common names, such as with fish. Jellyfish, crayfish, and silverfish all have the word "fish" in their name, but belong to three different species. In the eighteenth century, Carl Linnaeus invented a naming system for species that included using the Latin scientific name of a species, called the *binomial*, which has two parts: the *genus*, which comes first, and the *specific epithet*, which comes second. Similar species are grouped into the same genus. The Linnean system is the commonly used taxonomic system today and, moving from comprehensive similarities to more general similarities, classifies organisms into their species, genus, family, order, class, phylum, kingdom, and domain. *Homo sapiens* is the Latin scientific name for humans.

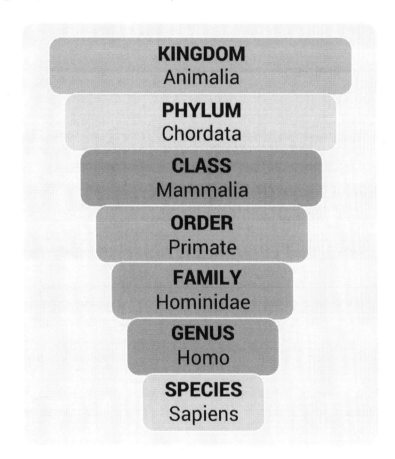

Phylogenetic trees are branching diagrams that represent the evolutionary history of a species. The branch points most often match the classification groups set forth by the Linnean system. Using this system helps elucidate the relationship between different groups of organisms. The diagram below is that of an empty phylogenetic tree:

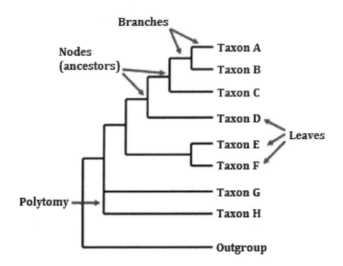

Each branch of the tree represents the divergence of two species from a common ancestor. For example, the coyote is known as Canis latrans and the gray wolf is known as Canis lupus. Their common ancestor, the Canis lepophagus, which is now extinct, is where their shared genus derived.

Characteristics of Bacteria, Animals, Plants, Fungi, and Protists
As discussed earlier, there are two distinct types of cells that make up most living organisms: prokaryotic and eukaryotic. Bacteria (and archaea) are classified as prokaryotic cells, whereas animal, plant, fungi, and protist cells are classified as eukaryotic cells.

Although animal cells and plant cells are both eukaryotic, they each have several distinguishing characteristics. *Animal cells* are surrounded by a plasma membrane, while *plant cells* have a cell wall made up of cellulose that provides more structure and an extra layer of protection for the cell. Animals use oxygen to breathe and give off carbon dioxide, while plants do the opposite—they take in carbon dioxide and give off oxygen. Plants also use light as a source of energy. Animals have highly developed sensory and nervous systems and the ability to move freely, while plants lack both abilities. Animals, however, cannot make their own food and must rely on their environment to provide sufficient nutrition, whereas plants do make their own food.

Fungal cells are typical eukaryotes, containing both a nucleus and membrane-bound organelles. They have a cell wall, similar to plant cells; however, they use oxygen as a source of energy and cannot perform photosynthesis. They also depend on outside sources for nutrition and cannot produce their own food. Of note, their cell walls contain *chitin.*

Protists are a group of diverse eukaryotic cells that are often grouped together because they do not fit into the categories of animal, plant, or fungal cells. They can be categorized into three broad categories: protozoa, protophyta, and molds. These three broad categories are essentially "animal - like," "plant-like," and "fungus-like," respectively. All of them are unicellular and do not form tissues. Besides this simple similarity, protists are a diverse group of organisms with different characteristics, life cycles, and cellular structures.

Major Structures of Plants and Their Functions

Characteristics of Vascular and Nonvascular Plants
Plants that have an extensive vascular transport system are called *vascular plants*. Those plants without a transport system are called *nonvascular plants*. Approximately ninety-three percent of plants that are currently living and reproducing are vascular plants. The cells that comprise the vascular tissue in vascular plants form tubes that transport water and nutrients through the entire plant. Nonvascular plants include mosses, liverworts, and hornworts. They do not retain any water; instead, they transport water using other specialized tissue. They have structures that look like leaves, but are actually just single sheets of cells without a cuticle or stomata.

Structure and Function of Roots, Leaves, and Stems
Roots are responsible for anchoring plants in the ground. They absorb water and nutrients and transport them up through the plant. *Leaves* are the main location of photosynthesis. They contain *stomata*, which are pores used for gas exchange, on their underside to take in carbon dioxide and release oxygen. *Stems* transport materials through the plant and support the plant's body. They contain *xylem*, which conducts water and dissolved nutrients upward through the plant, and *phloem*, which conducts sugars and metabolic products downward through the leaves.

Asexual and Sexual Reproduction
Plants can generate future generations through both asexual and sexual reproduction. Asexually, plants can go through an artificial reproductive technique called *budding*, in which parts from two or more plants of the same species are joined together with the hope that they will begin to grow as a single plant.

Sexual reproduction of flowers can happen in a couple of ways. *Angiosperms* are flowering plants that have seeds. The flowers have male parts that make pollen and female parts that contain ovules. Wind, insects, and other animals carry the pollen from the male part to the female part in a process called *pollination*. Once the ovules are pollinated, or fertilized, they develop into seeds that then develop into new plants. In many angiosperms, the flowers develop into fruit, such as oranges, or even hard nuts, which protect the seeds inside of them.

Nonvascular plants reproduce by sexual reproduction involving *spores*. Parent plants send out spores that contain a set of chromosomes. The spores develop into sperm or eggs, and fertilization is similar to that in humans. Sperm travel to the egg through water in the environment. An embryo forms and then a new plant grows from the embryo. Generally, this happens in damp places.

Growth
Germination is the process of a plant growing from a seed or spore, such as when a seedling sprouts from a seed or a sporeling grows from a spore. Plants then grow by *elongation*. Plant cell walls are modified by the hormone auxin, which allows for cell elongation. This process is regulated by light and phytohormones, which are plant hormones that regulate growth, so plants are often seen growing toward the sun.

Uptake and Transport of Nutrients and Water
Plant roots are responsible for bringing nutrients and water into the plant from the ground. The nutrients are not used as food for the plant, but rather to maintain the plant's health so that the plant can make its own food during photosynthesis. The xylem and phloem in the stem help with transport of water and other substances throughout the plant.

Responses to Stimuli

Because plants have limited mobility, they often respond to stimuli through changes in their growth behavior. *Tropism* is a response to stimuli that causes the plant to grow toward or away from the stimuli:

- *Phototropism*: A reaction to light that causes plants to grow toward the source of the light

- *Thermotropism*: A response to changes in temperature

- *Hydrotropism*: A response to a change in water concentration

- *Gravitropism*: A response to gravity that causes roots to follow the pull of gravity and grow downward, but also causes plant shoots to act against gravity and grow upward

Basic Anatomy and Physiology of Animals, Including the Human Body

Response to Stimuli and Homeostasis

A *stimulus* is a change in the environment, either internal or external, around an organism that is received by a sensory receptor and causes the organism to react. *Homeostasis* is the stable state of an organism. When an organism reacts to stimuli, it works to counteract the change in order to reach homeostasis again.

Exchange with the Environment

Animals exchange gases and nutrients with the environment through several different organ systems. The *respiratory system* mediates the exchange of gas between the air and the circulating blood, mainly through the act of breathing. It filters, warms, and humidifies the air that gets inhaled and then passes it into the blood stream. The main function of the *excretory system* is to eliminate excess material and fluids in the body. The kidneys and bladder work together to filter organic waste products, excess water, and electrolytes from the blood that are generated by the other physiologic systems, and excrete them from the body. The *digestive system* is a group of organs that work together to transform ingested food and liquid into energy, which can then be used by the body as fuel. Once all of the nutrients are absorbed, the waste products are excreted from the body.

Internal Transport and Exchange

The *circulatory system* is composed of the heart and blood vessels. The *heart* acts as a pump and works to circulate blood throughout the body. Blood circulates throughout the body in a system of vessels that includes arteries, veins, and capillaries. It distributes oxygen, nutrients, and hormones to all of the cells in the body. *Arteries* transport oxygen-rich blood from the heart to the rest of the tissues in the body. The largest artery is the *aorta*. *Veins* collect oxygen-depleted blood from tissues and organs and return it back to the heart. *Capillaries* are the smallest of the blood vessels and do not function individually. Instead, they work together in a unit—called a *capillary bed*—to transport both oxygen-rich and oxygen-poor blood to other vessels.

Control Systems

The *nervous system* is one of the smallest but most complex organ systems in the human body. It consists of all of the neural tissue and is in charge of controlling and adjusting the activities of all of the other systems of the body. It is divided into the *central nervous system (CNS)* and the *peripheral nervous system (PNS)*. The CNS is where intelligence, memory, learning, and emotions are processed. It is responsible for processing and coordinating sensory data and motor commands. The PNS is responsible for relaying sensory information and motor commands between the CNS and peripheral tissues and systems.

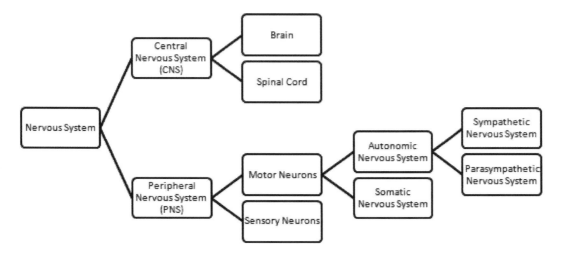

The *endocrine system* is made up of the ductless tissues and glands that secrete hormones into the *interstitial fluids* of the body, which are the fluids that surround tissue cells within the body. This system works closely with the nervous system to regulate the other physiologic systems in order to maintain homeostasis. It acts by releasing hormones into the bloodstream, which are then distributed to the whole body.

Movement and Support
The adult *skeletal system* consists of the 206 bones that make up the skeleton, as well as the cartilage, ligaments, and other connective tissues that stabilize the bones. It provides structural support for the entire body, a framework for the soft tissues and organs to attach to, and acts as a protective barrier for some organs, such as the ribs protecting the heart and lungs, and the vertebrae protecting the spinal cord.

The *muscular system* is responsible for all body movement that occurs. Body movements occur by muscle contractions that cause specific actions or joint movements. There are approximately seven hundred muscles in the body that are attached to the bones of the skeletal system. As mentioned, there are three types of muscle tissue: skeletal, cardiac, and smooth. *Skeletal muscles* are voluntary muscles that attach to bones through tendons. *Cardiac muscle tissue* is found only in the heart and is involuntary. *Smooth muscle tissue* lines the walls of hollow structures, such as blood vessels, the stomach, and the bladder, and is involuntary. When smooth muscle tissue contracts, the structure it lines narrows or constricts.

Reproduction and Development
The *reproductive system* is responsible for producing and maintaining functional reproductive cells in the human body. The human male and female reproductive systems are very different from each other. The male gonads, called *testes*, mainly secrete testosterone, which is a steroid hormone that controls the development and maintenance of male physical characteristics. They also produce *sperm cells*, which are responsible for fertilizing the female reproductive cell in order to produce offspring. The female gonads, called *ovaries*, generally produce one immature egg per month. They are also responsible for secreting the hormones estrogen and progesterone. Once an egg is fertilized by a sperm cell, an *embryo* develops. After approximately 40 weeks of gestation, a baby is born.

Immune System
The *immune system* comprises cells, tissues, and organs that work together to protect the body from harmful foreign invaders. It is important for the body's immune system to be able to distinguish between *pathogens*, such as viruses and parasites, and the body's own healthy tissue. There are two types of immune systems that work to defend the body against infection: the innate system and the adaptive system. The *innate immune system* works without having a memory of the pathogens it defended against previously. The *adaptive immune system* creates a memory of the pathogen that it fought against, so that the body can respond again in an efficient manner the next time the pathogen is encountered. When *antigens* or *allergens* such as pollen are encountered, antibodies are secreted to inactivate the antigen and protect the body.

If the immune system is not functioning properly, the body may develop an *autoimmune disorder*. In this case, the body cannot distinguish between itself and foreign pathogens, so it attacks itself unnecessarily.

Key Aspects of Ecology

Population Dynamics
Population dynamics is the study of the composition of populations, including size, age, and the biological and environmental processes that cause changes. These can include immigration, emigration, births, and deaths.

Growth Curves and Carrying Capacity
Population dynamics can be characterized by *growth curves*. Growth can either be *unrestricted*, which is modeled by an exponential curve, or *restricted*, which is modeled by a logistic curve. Population growth can be restricted by environmental factors such as the availability of food and water sources, habitat, and other necessities. The *carrying capacity* of a population is the maximum population size that an environment can sustain indefinitely, given all of the above factors.

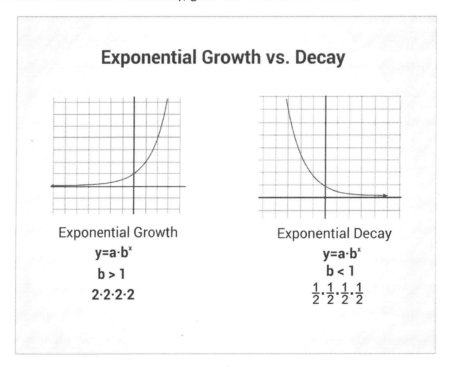

Exponential Growth vs. Decay

Exponential Growth
$y = a \cdot b^x$
$b > 1$
$2 \cdot 2 \cdot 2 \cdot 2$

Exponential Decay
$y = a \cdot b^x$
$b < 1$
$\frac{1}{2} \cdot \frac{1}{2} \cdot \frac{1}{2} \cdot \frac{1}{2}$

Behavior
Different species within a population can act differently regarding their environment. Some species display *territoriality*, which is a specific type of competition that excludes other species from a given area. It can be shown through specific animal calls, intimidating behavior, or marking an area with scents, and is often a display of defense.

Intraspecific Relationships
Intraspecific relationships is a term that describes the competition and cooperation between organisms that belong to the same species. They may compete for the same food sources, or for mates that are necessary for their personal survival and reproduction. Stronger organisms may display dominance that allows them to reside at the top of a social hierarchy and obtain better food and higher quality mates. However, organisms may also cooperate with each other in order to benefit the larger group; for instance, they may divide laborious activities among themselves.

Community Ecology
An *ecological community* is a group of species that interact and live in the same location. Because of their shared environment, they tend to have a large influence on each other.

Niche
An *ecological niche* is the role that a species plays in its environment, including how it finds its food and shelter. It could be a predator of a different species, or prey for a larger species.

Species Diversity
Species diversity is the number of different species that cohabitate in an ecological community. It has two different facets: *species richness*, which is the general number of species, and *species evenness*, which accounts for the population size of each species.

Interspecific Relationships
Interspecific relationships include the interactions between organisms of different species. The following list defines the common relationships that can occur:

- *Commensalism*: One organism benefits while the other is neither benefited nor harmed

- *Mutualism*: Both organisms benefit

- *Parasitism*: One organism benefits and the other is harmed

- *Competition*: Two or more species compete for limited resources that are necessary for their survival

- *Predation (Predator-Prey)*: One species is a food source for another species

Ecosystems
An *ecosystem* includes all of the living organisms and nonliving components of an environment (each community) and their interactions with each other.

Biomes

A *biome* is a group of plants and animals that are found in many different continents and have the same characteristics because of the similar climates in which they live. Each biome is composed of all of the ecosystems in that area. Five primary types of biomes are aquatic, deserts, forests, grasslands, and tundra. The sum total of all biomes comprises the Earth's biosphere.

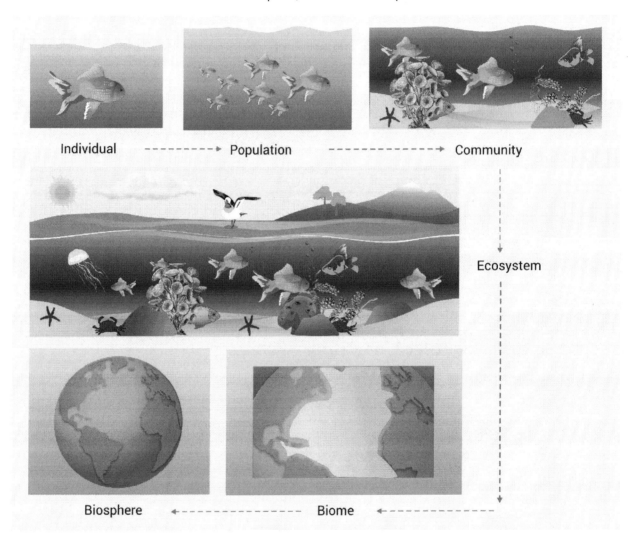

Stability and Disturbances

Ecological stability is the ability of an ecosystem to withstand changes that are occurring within it. With *regenerative stability*, an ecosystem may change, but then quickly return to its previous state. *Constant stability* occurs in ecosystems that remain unchanged despite the changes going on around them.

An *ecological disturbance* is a change in the environment that causes a larger change in the ecosystem. Smaller disturbances include fires and floods. Larger disturbances include the *climate change* that is currently occurring. Gas emissions from human activity are causing the atmosphere to warm up, which is changing the Earth's water systems and making weather more extreme. The increase in temperature is causing greater evaporation of the water sources on Earth, creating droughts and depleting natural water sources. This has also caused many of the Earth's glaciers to begin melting, which can change the salinity of the oceans.

Changes in the environment can cause an *ecological succession* to occur, which is the change in structure of the species that coexist in an ecological community. When the environment changes, resources available to the different species also change. For example, the formation of sand dunes or a forest fire would change the environment enough to allow a change in the social hierarchy of the coexisting species.

Energy Flow

Ecosystems are maintained by cycling the energy and nutrients that they obtain from external sources. The process can be diagramed in a *food web*, which represents the feeding relationship between species in a community. The different levels of the food web are called *trophic levels*. The first trophic level generally consists of plants, algae, and bacteria. The second trophic level consists of herbivores. The third trophic level consists of predators that eat herbivores. The trophic levels continue on to larger and larger predators. *Decomposers* are an important part of the food chain that are not at a specific trophic level. They eat decomposing things on the ground that other animals do not want to eat. This allows them to provide nutrients to their own predators.

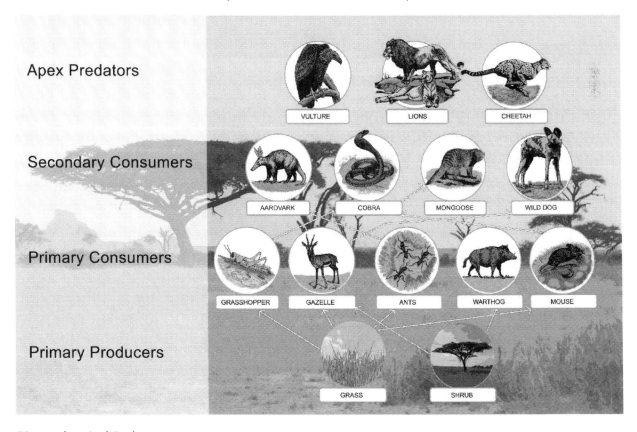

Biogeochemical Cycles

Biogeochemical cycles are the pathways by which chemicals move through the *biotic*, or biospheric, and *abiotic*, or atmospheric, parts of the Earth. The most important biogeochemical cycles include the water, carbon, and nitrogen cycles. *Water* goes through an evaporation, condensation, and precipitation cycle. *Nitrogen* makes up seventy-eight percent of the Earth's atmosphere and can affect the rate of many ecosystem processes, such as production of the primary producers at the first trophic level of the food web. The *carbon cycle* has many steps that are vitally important for sustaining life on Earth.

The water cycle:

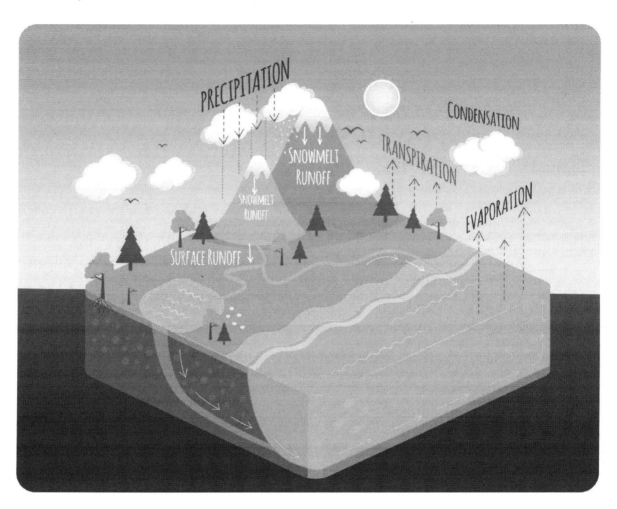

The nitrogen cycle:

The Nitrogen Cycle

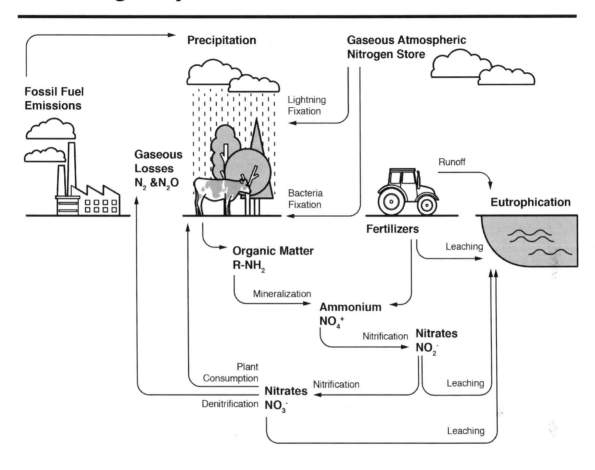

The Carbon Cycle

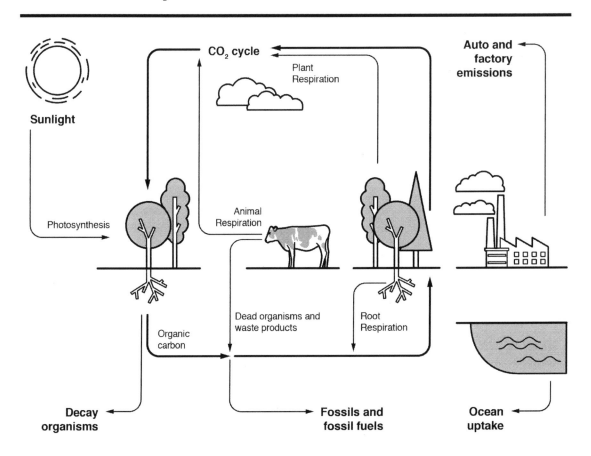

Practice Questions

1. What is the total mechanical energy of a system?
    a. The total potential energy
    b. The total kinetic energy
    c. Kinetic energy plus potential energy
    d. Kinetic energy minus potential energy
    e. Potential energy divided by kinetic energy

2. What does the Lewis Dot structure of an element represent?
    a. The outer proton valence shell population
    b. The inner electron valence shell population
    c. The positioning of the element's protons
    d. The positioning of the element's neutrons
    e. The outer electron valence shell population

3. What is the name of the scale used in sound level meters to measure the intensity of sound waves?
    a. Doppler
    b. Electron
    c. Watt
    d. Decibel
    e. Seismograph

4. Which statement is true regarding electrostatic charges?
    a. Like charges attract.
    b. Like charges repel.
    c. Like charges are neutral.
    d. Like charges neither attract nor repel.
    e. Like charges attack each other.

5. What is the name of this compound: CO?
    a. Carbonite oxide
    b. Carbonic dioxide
    c. Carbonic oxide
    d. Carbon monoxide
    e. Carbonite dioxide

6. What is the molarity of a solution made by dissolving 4.0 grams of $NaCl$ into enough water to make 120 mL of solution? The atomic mass of Na is 23.0 g/mol and Cl is 35.5 g/mol.
    a. 0.34 M
    b. 0.57 M
    c. 0.034 M
    d. 0.057 M
    e. 0.094 M

7. Considering a gas in a closed system, at a constant volume, what will happen to the temperature if the pressure is increased?

a. The temperature will stay the same

b. The temperature will decrease

c. The temperature will increase

d. It cannot be determined with the information given

e. The temperature will increase and then slowly decrease.

8. What is the current when a 3.0 V battery is wired across a lightbulb that has a resistance of 6.0 ohms?

a. 0.5 A

b. 18.0 A

c. 0.5 J

d. 18.0 J

e. 9.0 J

9. According to Newton's Three Laws of Motion, which of the following is true?

a. Two objects cannot exert a force on each other without touching.

b. An object at rest has no inertia.

c. The weight of an object is the same as the mass of the object.

d. Volume is equal to the change in momentum per change in time.

e. The weight of an object is equal to the mass of an object multiplied by gravity.

10. What is the chemical reaction when a compound is broken down into its elemental components called?

a. A synthesis reaction

b. A decomposition reaction

c. An organic reaction

d. An oxidation reaction

e. A composition reaction

11. What is ONE reason why speciation can occur?

a. Geographic separation

b. Seasons

c. Daylight

d. A virus

e. Gravity

12. What is the broadest, or LEAST specialized, classification of the Linnean taxonomic system?

a. Species

b. Family

c. Order

d. Phylum

e. Domain

13. How are fungi similar to plants?
    a. They have a cell wall
    b. They contain chloroplasts
    c. They perform photosynthesis
    d. They use carbon dioxide as a source of energy
    e. They use carbon monoxide as a source of energy

14. What important function are the roots of plants responsible for?
    a. Absorbing water from the surrounding environment
    b. Performing photosynthesis
    c. Conducting sugars downward through the leaves
    d. Supporting the plant body
    e. Seed development and reproduction

15. Which of the following would occur in response to a change in water concentration?
    a. Phototropism
    b. Thermotropism
    c. Gravitropism
    d. Hydrotropism
    e. Geotropism

16. What is the MAIN function of the respiratory system?
    a. To eliminate waste through the kidneys and bladder
    b. To exchange gas between the air and circulating blood
    c. To transform food and liquids into energy
    d. To excrete waste from the body
    e. To produce and secrete hormones

17. What type of vessel carries oxygen-rich blood from the heart to other tissues of the body?
    a. Veins
    b. Intestines
    c. Bronchioles
    d. Arteries
    e. Capillaries

18. Which system comprises the 206 bones of the body?
    a. Skeletal
    b. Muscular
    c. Endocrine
    d. Reproductive
    e. Lymphatic

19. Which factor is NOT a consideration in population dynamics?
    a. Size and age of population
    b. Immigration
    c. Hair color
    d. Number of births
    e. Number of deaths

20. Which type of diagram describes the cycling of energy and nutrients of an ecosystem?
    a. Food web
    b. Phylogenetic tree
    c. Fossil record
    d. Pedigree chart
    e. Photosynthesis

Answer Explanations

1. C: In any system, the total mechanical energy is the sum of the potential energy and the kinetic energy. Either value could be zero but it still must be included in the total. Choices *A* and *B* only give the total potential or kinetic energy, respectively. Choice *D* gives the difference in the kinetic and potential energy. Choice *E* gives the quotient of potential energy divided by kinetic energy, which is incorrect.

2. E: A Lewis Dot diagram shows the alignment of the valence (outer) shell electrons and how readily they can pair or bond with the valence shell electrons of other atoms to form a compound. Choice *B* is incorrect because the Lewis Dot structure aids in understanding how likely an atom is to bond or not bond with another atom, so the inner shell would add no relevance to understanding this likelihood. The positioning of protons and neutrons concerns the nucleus of the atom, which again would not lend information to the likelihood of bonding.

3. D: The decibel scale is used to measure the intensity of sound waves. The decibel scale is a ratio of a particular sound's intensity to a standard value. Since it is a logarithmic scale, it is measured by a factor of 10. Choice *A* is the name of the effect experienced by an observer of a moving wave; Choice *B* is a particle in an atom; Choice *C* is a unit for measuring power; and Choice *E* is a device used to measure an earthquake.

4. B: For charges, *like charges repel* each other and *opposite charges attract* each other. Negatives and positives will attract, while two positive charges or two negative charges will repel each other. Charges have an effect on each other, so Choices *C* and *D* are incorrect.

5. D: The naming of compounds focuses on the second element in a chemical compound. Elements from the non-metal category are written with an "ide" at the end. The compound CO has one carbon and one oxygen, so it is called carbon monoxide. Choices *B* and *E* represent that there are two oxygen atoms, and Choices *A*, *B*, and *E* incorrectly alter the name of the first element, which should remain as carbon.

6. B: To solve this, the number of moles of NaCl needs to be calculated:

First, to find the mass of NaCl, the mass of each of the molecule's atoms is added together as follows:

$$23.0\text{g (Na)} + 35.5\text{g (Cl)} = 58.8\text{g NaCl}$$

Next, the given mass of the substance is multiplied by one mole per total mass of the substance:

$$4.0\text{g NaCl} \times (1 \text{ mol NaCl}/58.5\text{g NaCl}) = 0.068 \text{ mol NaCl}$$

Finally, the moles are divided by the number of liters of the solution to find the molarity:

$$(0.068 \text{ mol NaCl})/(0.120\text{L}) = 0.57 \text{ M NaCl}$$

Choice *A* incorporates a miscalculation for the molar mass of NaCl, and Choices *C* and *D* both incorporate a miscalculation by not converting mL into liters (L), so they are incorrect by a factor of 10. Choice *E* is incorrect.

7. C: According to the *ideal gas law* ($PV = nRT$), if volume is constant, the temperature is directly related to the pressure in a system. Therefore, if the pressure increases, the temperature will increase in direct proportion. Choice *A* would not be possible, since the system is closed and a change is occurring, so the temperature will change. Choice *B* incorrectly exhibits an inverse relationship between pressure and temperature, or $P = 1/T$. Choice *D* is incorrect because even without actual values for the variables, the relationship and proportions can be determined. Choice *E* is incorrect; the temperature will slowly increase as the pressure increases.

8. A: According to Ohm's Law: $V = IR$, so using the given variables: $3.0\,V = I \times 6.0\,\Omega$

Solving for I: $I = 3.0\,V/6.0\,\Omega = 0.5\,A$

Choice *B* incorporates a miscalculation in the equation by multiplying 3.0 V by 6.0 Ω, rather than dividing these values. Choices *C, D* and *E* are labeled with the wrong units; Joules measure energy, not current.

9. E: The weight of an object is equal to the mass of the object multiplied by gravity. According to Newton's Second Law of Motion, $F = m \times a$. Weight is the force resulting from a given situation, so the mass of the object needs to be multiplied by the acceleration of gravity on Earth: $W = m \times g$. Choice *A* is incorrect because, according to Newton's first law, all objects exert some force on each other, based on their distance from each other and their masses. This is seen in planets, which affect each other's paths and those of their moons. Choice *B* is incorrect because an object in motion or at rest can have inertia; inertia is the resistance of a physical object to change its state of motion. Choice *C* is incorrect because the mass of an object is a measurement of how much substance of there is to the object, while the weight is gravity's effect of the mass. Choice *D* is incorrect; force is equal to the change in momentum per change in time, not volume.

10. B: A decomposition reaction breaks down a compound into its constituent elemental components. Choice *A* is incorrect because a synthesis reaction joins two or more elements into a single compound. Choice *C*, an organic reaction, is not possible, since it needs carbon and hydrogen for a reaction. Choice *D*, oxidation/reduction (redox or half) reaction, is incorrect because it involves the loss of electrons from one species (oxidation) and the gain of electrons to the other species (reduction). There is no notation of this occurring within the given reaction, so it is not correct. Choice *E* is incorrect; A composition reaction produces a single substance from multiple reactants, opposed to producing multiple products from a single reactant.

11. A: Speciation is the method by which one species splits into two or more species. In allopatric speciation, one population is divided into two subpopulations. If a drought occurs and a large lake becomes divided into two smaller lakes, each lake is left with its own population that cannot intermingle with the population of the other lake. When the genes of these two subpopulations are no longer mixing with each other, new mutations can arise and natural selection can take place.

12. E: In the Linnean system, organisms are classified as follows, moving from comprehensive and specific similarities to fewer and more general similarities: species, genus, family, order, class, phylum, kingdom, and domain. A popular mnemonic device to remember the Linnean system is "Dear King Philip came over for good soup."

13. A: Fungal cells have a cell wall, similar to plant cells; however, they use oxygen as a source of energy and cannot perform photosynthesis. Because they do not perform photosynthesis, fungal cells do not contain chloroplasts.

14. A: Roots are responsible for absorbing water and nutrients that will get transported up through the plant. They also anchor the plant to the ground. Photosynthesis occurs in leaves, stems transport materials through the plant and support the plant body, and phloem moves sugars downward to the leaves. Choice *E* is incorrect because flowers are responsible for seed development and reproduction, not the roots.

15. D: Tropism is a response to stimuli that causes the plant to grow toward or away from the stimuli. Hydrotropism is a response to a change in water concentration. Phototropism is a reaction to light that causes plants to grow toward the source of the light. Thermotropism is a response to changes in temperature. Gravitropism is a response to gravity that causes roots to follow the pull of gravity and grow downward, but also causes plant shoots to act against gravity and grow upward. Geotropism causes some plants to grow upward and some plants to grow toward the ground; it is the growth of plants with respect to gravity.

16. B: The respiratory system mediates the exchange of gas between the air and the circulating blood, mainly by the act of breathing. It filters, warms, and humidifies the air that gets breathed in and then passes it into the blood stream. The digestive system transforms food and liquids into energy and helps excrete waste from the body. Eliminating waste via the kidneys and bladder is a function of the urinary system. Glands that produce and secrete hormones are located in the endocrine system.

17. D: Arteries carry oxygen-rich blood from the heart to the other tissues of the body. Veins carry oxygen-poor blood back to the heart. Intestines carry digested food through the body. Bronchioles are passageways that carry air from the nose and mouth to the lungs. Capillaries exchange nutrients, waste, and oxygen with tissues at the cellular level.

18. A: The skeletal system consists of the 206 bones that make up the skeleton, as well as the cartilage, ligaments, and other connective tissues that stabilize the bones. The skeletal system provides structural support for the entire body, a framework for the soft tissues and organs to attach to, and acts as a protective barrier for some organs, such as the ribs protecting the heart and lungs, and the vertebrae protecting the spinal cord. The muscular system includes skeletal muscles, cardiac muscle, and the smooth muscles found on the inside of blood vessels. The endocrine system uses ductless glands to produce hormones that help maintain hemostasis, the reproductive system is responsible for the productions of egg and sperm cells, and the lymphatic system is a network of lymphatic vessels that carry a clear fluid called lymph towards the heart.

19. C: Population dynamics looks at the composition of populations, including size and age, and the biological and environmental processes that cause changes. These can include immigration, emigration, births, and deaths.

20. A: Ecosystems are maintained by cycling the energy and nutrients that they obtain from external sources. The process can be diagramed in a food web, which represents the feeding relationship between the species in a community. A phylogenetic tree shows inferred evolutionary relationships among species and is similar to the fossil record. A pedigree chart shows occurrences of phenotypes of a particular gene through the generations of an organism. A photosynthesis diagram shows a process by which plants make their own food.

Table Reading

This section of the test measures whether you can read a table accurately. The table below shows X- and Y-values in a grid format. The numbers on the horizontal, or top, part of the table are the X-values, and the numbers on the vertical, or side, part of the table are the Y-values. In other words, the X-values correspond to columns, and the Y-values correspond to rows. Each question will provide you with two coordinates, an X and a Y, and you must quickly interpret the table to determine the correct value. Each question will have five possible answers, and you should choose the correct one.

Using the example table below, here are some practice questions and explanations to better help you understand how to use these tables.

	-3	-2	-1	0	+1	+2	+3
+3	14	15	17	19	20	21	22
+2	15	17	19	21	22	23	24
+1	16	18	20	22	24	25	26
0	18	19	21	23	25	26	28
-1	20	22	23	25	27	28	30
-2	21	23	24	26	28	29	31
-3	22	24	25	27	29	30	32

Example Problem 1

X	Y		A	B	C	D	E
0	+1		22	25	20	19	16

The answer to this example question would be A, 22. To answer this problem, first go to the X-axis and find the 0 column. Then follow the 0 column down to the +1 row on the Y-axis. The number at that point is 22. Some corresponding coordinates for the incorrect answers in this example would be:

B = 25: +1, 0

C = 20: +1, +3

D = 19: 0, +3

E = 16: -3, +1

What are some helpful tips for answering these questions correctly?

- Take your time. The easiest way to make mistakes in this section is assuming that the questions are easy and subsequently making simple mistakes.

- Make sure to correctly read the X and Y headings. It is easy to select the second column when you have an X-value of 2, but that wouldn't be the right column.

- Start with the X coordinate and stay within that selected column. Do not mistakenly travel to another column when you are selecting the Y-value.

- After you have the X column selected, go down to the corresponding Y-value row. Remember, don't change columns!

- Use your fingers or a piece of paper to mark the columns and rows as it helps you keep track of where you are and not jump to a neighboring column or row.

- Practice these types of questions until you are comfortable answering them correctly and within a certain window of time. On test day, take the same amount of time as you had been with the practice questions.

Practice Questions

The next five questions are based on the following table.

	-3	-2	-1	0	1	2	3
-3	20	21	24	27	23	26	25
-2	21	22	29	32	31	30	26
-1	23	25	26	27	24	21	20
0	25	26	30	31	32	29	22
1	31	28	29	22	20	24	27
2	28	33	20	30	21	23	25
3	30	28	26	24	22	20	32

1. (-3, 3)
   - a. 30
   - b. 20
   - c. 32
   - d. 23
   - e. 24

2. (0, -2)
   - a. 30
   - b. 26
   - c. 32
   - d. 21
   - e. 29

3. (+1, 0)
   - a. 22
   - b. 30
   - c. 24
   - d. 27
   - e. 32

4. (-2, -3)
   - a. 23
   - b. 21
   - c. 33
   - d. 26
   - e. 20

5. (-1, 0)
   - a. 22
   - b. 27
   - c. 24
   - d. 30
   - e. 23

The next five questions are based on the following table.

	-3	-2	-1	0	1	2	3
-3	10	11	13	15	18	17	19
-2	17	20	19	16	14	12	10
-1	18	19	17	20	13	11	16
0	13	22	18	14	12	10	11
1	11	19	20	16	15	14	18
2	19	23	22	12	17	15	20
3	14	15	12	18	11	22	21

6. (3, 3)
   a. 10
   b. 21
   c. 16
   d. 12
   e. 19

7. (-2, 1)
   a. 14
   b. 11
   c. 20
   d. 17
   e. 19

8. (0, 0)
   a. 14
   b. 15
   c. 13
   d. 10
   e. 12

9. (1, -2)
   a. 19
   b. 17
   c. 22
   d. 14
   e. 11

10. (-1, 0)
   a. 20
   b. 12
   c. 15
   d. 13
   e. 18

The next five questions are based on the following table.

	-3	-2	-1	0	1	2	3
-3	30	31	32	33	34	35	36
-2	33	34	35	36	37	38	39
-1	34	40	41	39	38	37	36
0	35	33	31	32	30	40	41
1	31	35	37	33	39	41	43
2	37	36	33	34	35	39	38
3	32	30	31	35	36	34	31

11. (2, -2)
    a. 34
    b. 36
    c. 39
    d. 35
    e. 38

12. (1, -3)
    a. 31
    b. 38
    c. 34
    d. 33
    e. 36

13. (3, 2)
    a. 38
    b. 34
    c. 37
    d. 39
    e. 33

14. (0, -1)
    a. 39
    b. 31
    c. 33
    d. 32
    e. 34

15. (-2, -1)
    a. 41
    b. 40
    c. 35
    d. 37
    e. 31

The next five questions are based on the following table.

	-3	-2	-1	0	1	2	3
-3	50	51	52	53	54	55	56
-2	53	55	56	57	58	59	52
-1	54	60	57	58	52	61	53
0	55	53	58	59	60	52	54
1	56	59	54	60	51	53	55
2	57	54	55	61	59	57	56
3	51	52	53	54	55	56	57

16. (-3, 2)
   a. 55
   b. 52
   c. 53
   d. 57
   e. 56

17. (-2, 0)
   a. 57
   b. 52
   c. 61
   d. 53
   e. 51

18. (1, 1)
   a. 51
   b. 57
   c. 52
   d. 54
   e. 55

19. (2, -1)
   a. 56
   b. 60
   c. 61
   d. 53
   e. 59

20. (-1, 3)
   a. 55
   b. 53
   c. 52
   d. 54
   e. 57

The next five questions are based on the following table.

	-3	-2	-1	0	1	2	3
-3	40	41	42	43	44	45	46
-2	42	44	43	54	47	46	45
-1	44	46	45	52	53	48	43
0	46	48	47	50	49	44	51
1	48	50	49	48	51	52	47
2	50	52	51	46	47	42	49
3	52	54	53	44	45	50	41

21. (0, 2)
   a. 46
   b. 44
   c. 54
   d. 48
   e. 52

22. (3, -2)
   a. 54
   b. 50
   c. 45
   d. 49
   e. 42

23. (-1, 1)
   a. 53
   b. 45
   c. 51
   d. 49
   e. 42

24. (2, 3)
   a. 49
   b. 54
   c. 41
   d. 45
   e. 50

25. (-2, 1)
   a. 47
   b. 50
   c. 48
   d. 46
   e. 52

The next five questions are based on the following table.

	-3	-2	-1	0	1	2	3
-3	5	6	7	8	9	10	11
-2	6	7	8	9	10	11	12
-1	8	9	10	11	12	13	14
0	10	11	12	13	14	15	16
1	12	13	14	15	16	17	6
2	14	15	16	17	6	5	8
3	16	17	6	5	8	7	10

26. (3, 0)
    a. 5
    b. 10
    c. 8
    d. 11
    e. 16

27. (1, 2)
    a. 17
    b. 6
    c. 16
    d. 10
    e. 8

28. (-1, -1)
    a. 16
    b. 14
    c. 10
    d. 12
    e. 7

29. (-3, -2)
    a. 7
    b. 12
    c. 6
    d. 14
    e. 8

30. (3, -1)
    a. 14
    b. 7
    c. 6
    d. 12
    e. 8

The next five questions are based on the following table.

	-3	-2	-1	0	1	2	3
-3	11	12	13	14	15	16	17
-2	12	13	14	15	16	17	18
-1	14	15	16	17	18	19	20
0	16	17	18	19	20	21	22
1	18	19	20	21	22	21	12
2	20	21	22	11	12	13	14
3	22	11	12	13	14	15	16

31. (-3, -1)
    a. 14
    b. 13
    c. 20
    d. 18
    e. 12

32. (0, 1)
    a. 20
    b. 17
    c. 21
    d. 18
    e. 14

33. (2, 1)
    a. 12
    b. 19
    c. 17
    d. 15
    e. 21

34. (3, -3)
    a. 22
    b. 17
    c. 16
    d. 11
    e. 20

35. (-1, 2)
    a. 19
    b. 16
    c. 14
    d. 22
    e. 12

The next five questions are based on the following table.

	-3	-2	-1	0	1	2	3
-3	20	21	22	23	24	25	26
-2	21	22	23	24	25	26	27
-1	23	24	25	26	27	28	29
0	25	26	27	28	29	30	31
1	27	28	29	30	31	20	21
2	29	30	31	20	21	22	23
3	31	20	21	22	23	24	25

36. (2, 2)
    a. 30
    b. 22
    c. 26
    d. 21
    e. 20

37. (1, -3)
    a. 27
    b. 23
    c. 24
    d. 29
    e. 22

38. (1, -1)
    a. 31
    b. 29
    c. 25
    d. 27
    e. 24

39. (2, -2)
    a. 26
    b. 30
    c. 22
    d. 20
    e. 21

40. (0, 3)
    a. 31
    b. 22
    c. 23
    d. 25
    e. 26

Answer Explanations

1. A: Select -3 on the horizontal X axis of the table and move down to +3 on the vertical Y axis. This number is 30. (-3, -3) is 20, (3, 3) is 32, (1, -3) is 23, and (-1, -3) is 24.

2. C: The X axis is 0, while the Y axis is -2; this number is 32. (0, 2) is 30, (-2, 0) is 26, (-2, -3) is 21, and (2, 0) is 29.

3. E: The X axis is +1, while the Y axis is 0; this number is 32. (0, 1) is 22, (-1, 0) is 30, (-1, -3) is 24, and (0, -1) is 27.

4. B: The X value is -2 and the Y value is -3; this number is 21. (-3, -1) is 23, (-2, -2) is 33, (2, -3) is 26, and (3, -1) is 20.

5. D: The X value is -1 and the Y value is 0; this number is 30. (0, 1) is 22, (0, -1) is 27, (-1, -3) is 24, and (-3, -1) is 23.

6. B: The X value is 3 and the Y value is 3; this number is 21. (-3, -3) is 10, (3, -1) is 16, (-1, 3) is 12, and (3, -3) is 19.

7. E: The X value is -2 and the Y value is 1; this number is 19. (1, -2) is 14, (2, -1) is 11, (-2,-2) is 20, and (-1, -1) is 17.

8. A: The X value is 0 and the Y value is 0; this number is 14. (0, -3) is 15, (-3, 0) is 13, (-3, -3) is 10, and (1, 0) is 12.

9. D: The X value is 1 and the Y value is -2; this number is 14. (-2, 1) is 19, (1, 2) is 17, (-1, 2) is 22, and (2, -1) is 11.

10. E: The X value is -1 and the Y value is 0; this number is 18. (0, -1) is 20, (1, 0) is 12, (0, -3) is 15, (-1, -3) is 13, and (-1, 0) is 18.

11. E: The X value is 2 and the Y value is -2; this number is 38. (-2, -2) is 34, (-2, 2) is 36, (2, 2) is 39, and (2, -3) is 35.

12. C: The X value is 1 and the Y value is -3; this number is 34. (-3, 1) is 31, (1, -1) is 38, (-1, 2) is 33, and (1, 3) is 36.

13. A: The X value is 3 and the Y value is 2; this number is is 38. (2, 3) is 34, (-3, 2) is 37, (3, -2) is 39, and (-3, -2) is 33.

14. A: The X value is 0 and the Y value is -1; this number is 39. (-1, 0) is 31, (0, -3) is 33, (-1, -3) is 32, and (0,2) is 34.

15. B: The X value is -2 and the Y value is -1; this number is 40. (2, 1) is 41, (-2, 1) is 35, (2, -1) is 37, and (-2, -3) is 31.

16. D: The X value is -3 and the Y value is 2; this number is 57. (2, -3) is 55, (3, -2) is 52, (-3, -2) is 53, and (2, 3) is 56.

17. D: The X value is -2 and the Y value is 0; this number is is 53. (0, -2) is 57, (2, 0) is 52, (0, 2) is 61, and (-2, -3) is 51.

18. A: The X value is 1 and the Y value is 1; this number is is 51. (-1, -1) is 57, (1, -1) is 52, (-1, 1) is 54, and (1, 3) is 55.

19. C: The X value is 2 and the Y value is -1; this number is 61. (-1, 2) is 56, (-2, -1) is 60, (2, 1) = 53, and (-2, 1) is 59.

20. B: The X value is -1 and the Y value is 3; this number is 53. (3, 1) is 55, (1, -1) is 52, (1, -3) is 54, and (-1, -1) is 57.

21. A: The X value is 0 and the Y value is 2; this number is 46. (2, 0) is 44, (0, -2) is 54, (-2, 0) is 48, and (0, -1) is 52.

22. C: The X value is 3 and the Y value is -2; this number is 45. (-2, 3) is 54, (-3, 2) is 50, (3, 2) is 49, and (-3, -2) is 42.

23. D: The X value is -1 and the Y value is 1; this number is 49. (1, -1) is 53, (-1, -1) is 45, (1, 1) is 51, and (-1, -3) is 42.

24. E: The X value is 2 and the Y value is 3; this number is 50. (3, 2) is 49, (-2, 3) is 54, (-2, -3) is 41, and (2, -3) is 45.

25. B: The X value is -2 and the Y value is 1; this number is 50. (1, -2) is 47, (2, -1) is 48, (-2, -1) is 46, and (2, 1) is 52.

26. E: The X value is 3 and the Y value is 0; this number is 16. (0,3) is 5, (-3, 0) is 10, (0, -3) is 8, and (3, -3) is 11.

27. B: The X value is 1 and the Y value is 2; this number is 6. (2, 1) is 17, (-1, 2) is 16, (1, -2) is 10, and (-1, -2) is 8.

28. C: The X value is -1 and the Y value is -1; this number is 10. (1, 1) is 16, (-1, 1) is 14, (1, -1) is 12, and (-1, -3) is 7.

29. C: The X value is -3 and the Y value is -2; this number is 6. (-2, -2) is 7, (3, -2) is 12, (-3, 2) is 14, and (3, 2) is 8.

30. A: The X value is 3 and the Y value is -1; this number is 14. (-1, 3) is 7, (3, 1) is 6, (-3, 1) is 12, and (-3, -1) is 8.

31. A: The X value is -3 and the Y value is -1; this number is 14. (-1, -3) is 13, (3, -1) is 20, (-3, 1) is 18, and (3, 1) = 12.

32. C: The X value is 0 and the Y value is 1; this number is 21. (1, 0) is 20, (0, -1) is 17, (-1, 0) is 18, and (0, -3) is 14.

33. E: The X value is 2 and the Y value is 1; this number is 21. (1, 2) is 12, (-2, 1) is 19, (2, -2) is 17, and (-2, -1) is 15.

34. B: The X value is 3 and the Y value is -3; this number is 17. (-3, 3) is 22, (3, 3) is 16, (-3, -3) is 11, and (3, -1) is 20.

35. D: The X value is -1 and the Y value is 2; this number is 22. (2, -1) is 19, (1, -2) is 16, (-1, -2) is 14, and (1, 2) is 12.

36. B: The X value is 2 and the Y value is 2; this number is 22. (-2, 2) is 30, (2, -2) is 26, (-2, -3) is 21, and (2, 1) is 20.

37. C: The X value is 1 and the Y value is -3; this number is 24. (-3, 1) is 27, (1, 3) is 23, (3, -1) is 29, and (-1, -3) is 22.

38. D: The X value is 1 and the Y value is -1; this number is 27. (1, 1) is 31, (-1, 1) is 29, (-1, -1) is 25, and (1, -3) is 24.

39. A: The X value is 2 and the Y value is -2; this number is 26. (-2, 2) is 30, (-2, -2) is 22, (2, 1) is 20, and (-2, -3) is 21.

40. B: The X value is 0 and the Y value is 3; this number is 22. (3, 0) is 31, (0, -3) is 23, (-3, 0) is 25, and (0, -1) is 26.

Block Counting

This section of the test measures your spatial reasoning and logic. You will be shown a three-dimensional (3-D) drawing of blocks. Some of these blocks may be touching, and, if so, you will be asked to determine the number of blocks that one particular block is touching. Keep in mind that some of the blocks may not be visible. There is no prior knowledge of block counting that is needed for this section of the test; just use logic and spatial reasoning.

In order for a block to "touch" another block, their faces must touch. If only two blocks' corners touch and not their faces, then they are not considered to be touching.

See the example problem below.

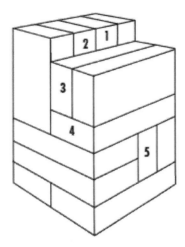

In this example, you know that Block 1 is touching 5 other blocks. Block 1 is touching Block 2 and 3, the block to the right of Block 1, Block 4, and Block 5.

Block 2 is also touching 5 blocks: Block 1, the block to the left of Block 2, Block 3, Block 4, and one of the horizontal blocks beside Block 5.

What about Block 4? Block 4 is touching an astonishing 9 blocks: Block 3, the block to the right of Block 3, Blocks 1 and 2 and the two on either side of them, Block 5 and the two blocks on the left of it, and the one block on the right of Block 5.

Sometimes you will have to infer how big a block is when only part of the block is shown.

Practice Questions

For questions 1-30, determine how many blocks the given block is touching.

Use the figure below for questions 1-5.

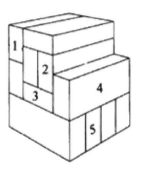

1. Block 1
    a. 1
    b. 2
    c. 3
    d. 4
    e. 5

2. Block 2
    a. 4
    b. 5
    c. 6
    d. 7
    e. 8

3. Block 3
    a. 5
    b. 6
    c. 7
    d. 8
    e. 9

4. Block 4
    a. 4
    b. 5
    c. 6
    d. 7
    e. 8

5. Block 5
    a. 2
    b. 3
    c. 4
    d. 5
    e. 6

Use the figure below for questions 6-10.

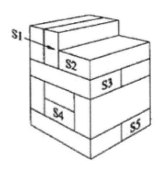

6. Block S1
    a. 1
    b. 2
    c. 3
    d. 4
    e. 5

7. Block S2
    a. 2
    b. 3
    c. 4
    d. 5
    e. 6

8. Block S3
    a. 5
    b. 6
    c. 7
    d. 8
    e. 9

9. Block S4
    a. 4
    b. 5
    c. 6
    d. 7
    e. 8

10. Block S5
    a. 3
    b. 4
    c. 5
    d. 6
    e. 7

Use the figure below for questions 11-17.

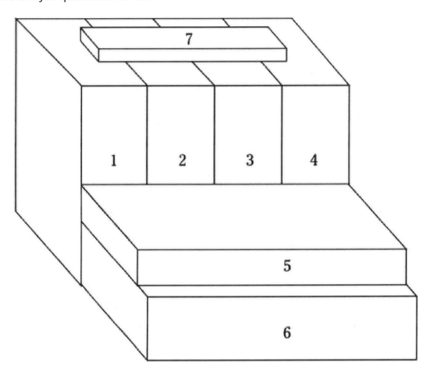

11. Block 1
    a. 1
    b. 2
    c. 3
    d. 4
    e. 5

12. Block 2
    a. 4
    b. 5
    c. 6
    d. 7
    e. 8

13. Block 3
    a. 5
    b. 6
    c. 7
    d. 2
    e. 1

14. Block 4
    a. 4
    b. 5
    c. 6
    d. 7
    e. 2

15. Block 5
    a. 2
    b. 3
    c. 4
    d. 5
    e. 6

16. Block 6
    a. 1
    b. 2
    c. 3
    d. 4
    e. 5

17. Block 7
    a. 3
    b. 4
    c. 5
    d. 6
    e. 7

Use the figure below for questions 18-22.

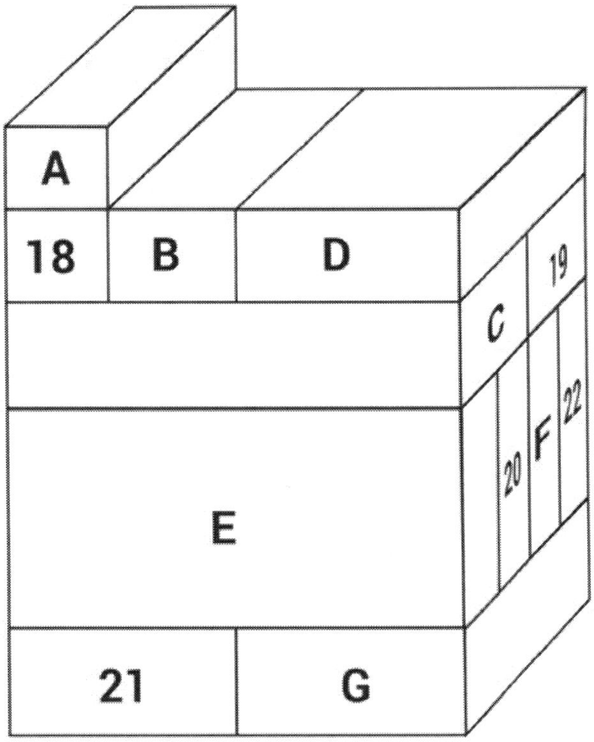

18. Block 18
    a. 1
    b. 2
    c. 3
    d. 4
    e. 5

19. Block 19
    a. 2
    b. 3
    c. 4
    d. 5
    e. 6

175

20. Block 20
    a. 1
    b. 2
    c. 3
    d. 4
    e. 5

21. Block 21
    a. 5
    b. 6
    c. 7
    d. 8
    e. 9

22. Block 22
    a. 4
    b. 5
    c. 6
    d. 7
    e. 8

Use the figure below for questions 23-26.

23. Block 1
    a. 1
    b. 2
    c. 3
    d. 4
    e. 5

24. Block 2
    a. 5
    b. 6
    c. 7
    d. 8
    e. 9

25. Block 3
    a. 4
    b. 5
    c. 6
    d. 7
    e. 8

26. Block 4
    a. 2
    b. 3
    c. 4
    d. 5
    e. 6

Use the figure below for questions 27-30.

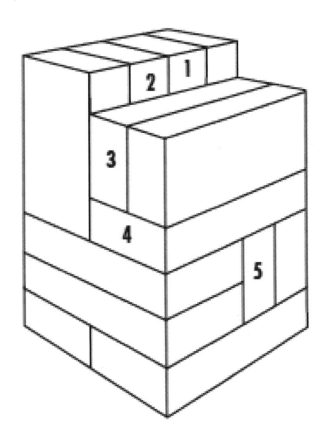

27. Block 1
    a. 1
    b. 2
    c. 3
    d. 4
    e. 5

28. Block 2
    a. 4
    b. 5
    c. 6
    d. 7
    e. 8

29. Block 3
    a. 2
    b. 3
    c. 4
    d. 5
    e. 6

30. Block 4
    a. 6
    b. 7
    c. 8
    d. 9
    e. 10

Answer Explanations

1. C: Block 1 touches 3 blocks: block below Block 1 and 2 blocks on the right side.

2. A: Block 2 touches 4 blocks: block above Block 2, block on the left of Block 2, Block 3, and Block 4.

3. D: Block 3 touches 8 blocks: 4 blocks on the bottom (including Block 5), Block 4, Block 2, block on the left of Block 2, and block below Block 1.

4. C: Block 4 touches 6 blocks: 4 blocks below (including Block 5), Block 2, and Block 3.

5. D: Block 5 touches 5 blocks: block on the left of Block 5, block on the right of Block 5, Block 4, Block 3, and block below Block 1.

6. D: Block S1 touches 4 blocks: block on the left of Block S1, Block S2, Block S3, and block on the right of Block S3.

7. B: Block S2 touches 3 blocks: Block S1, Block S3, and block on the right of Block S3.

8. C: Block S3 touches 7 blocks: block on the left of Block S1, Block S1, Block S2, block on the right of Block S3, block above Block S4, block on the left of Block S4, and block on the right of Block S4.

9. B: Block S4 touches 5 blocks: Block S5, block on the left of Block S5, block on the left of Block S4, block on right of Block S4, and block above Block S4.

10. B: Block S5 touches 4 blocks: block on the left of Block S5, block above Block S5, Block S4, and block on the left of Block S4.

11. D: Block 1 touches 4 blocks: Blocks 2, 5, 6, and 7.

12. B: Block 2 touches 5 blocks: Blocks 1, 3, 5, 6, and 7.

13. A: Block 3 touches 5 blocks: Blocks 2, 4, 5, 6, and 7.

14. A: Block 4 touches 4 blocks: Blocks 3, 5, 6, and 7.

15. D: Block 5 touches 5 blocks: Blocks 1, 2, 3, 4, and 6.

16. E: Block 6 touches 5 blocks: Blocks 1, 2, 3, 4, and 5.

17. B: Block 7 touches 4 blocks: Blocks 1, 2, 3, and 4.

18. D: Block 18 touches 4 blocks: Blocks A, B, C, and 19.

19. E: Block 19 touches 6 blocks: Blocks D, C, B, 18, 22, and F.

20. E: Block 20 touches 5 blocks: Blocks C, E,F, G, and 21.

21. A: Block 21 touches 5 blocks: Blocks E, G, 20, F, and 22.

22. A: Block 22 touches 4 blocks: Blocks 19, F, G, and 21.

23. C: Block 1 touches 3 blocks: block above Block 1, Block 3, and block to the left of Block 3.

24. C: Block 2 touches 7 blocks: block to the left of Block 5, Block 5, block to the left of Block 3, Block 3, 2 blocks to the back and left of Block 2, and Block 4.

25. D: Block 3 touches 7 blocks: Block 1, block above Block 1, block to the left of Block 3, Block 2, Block 4, and 2 blocks to the back and left of Block 2.

26. D: Block 4 touches 5 blocks: block to the left of Block 5, Block 5, block to the left of Block 3, Block 3, and Block 2.

27. E: Block 1 touches 5 blocks: Block 2, block to the right of Block 1, Block 3, Block 4, and Block 5.

28. B: Block 2 touches 5 blocks: block to the left of Block 2, Block 1, Block 3, Block 4, and the block below Block 4 and to the left of Block 5.

29. E: Block 3 touches 6 blocks: block to the left of Block 2, Block 2, Block 1, block to the right of Block 1, block to the right of Block 3, and Block 4.

30. D: Block 4 touches 9 blocks: block to the left of Block 2, Block 2, Block 1, block to the right of Block 1, Block 3, block to the right of Block 3, block below Block 4 and to the left of Block 5, Block 5, and block to the right of Block 5.

Instrument Comprehension

Purpose of Instrument-Comprehension Questions

The purpose of the instrument-comprehension questions is to ensure that vital instruments are understood. There needs to be a familiarity with the instruments and knowledge of their purposes, as well as how to read them.

Presentation of the Instrument-Comprehension Questions

Depending on the types of instruments covered in the test, the comprehension questions will vary in layout. For example, the practice questions in this section pertain to the artificial horizon and the compass. These two instruments are presented with the readings that you would need to be able to translate in order to determine the orientation of the aircraft, were they presented to you in the cockpit.

Reading the Compass and Artificial-Horizon Instruments

The compass on an aircraft is a navigational instrument used to determine the direction of flight, or the "aircraft heading." Compass reading is simple. As most people are aware, the needle reflects the direction being faced. When the aircraft is facing north, the needle will be pointing at the "N." If the aircraft is facing southwest, the needle will point to "SW."

The artificial horizon, also known as the attitude indicator, is an aircraft instrument that shows the aircraft orientation as compared with the Earth's horizon. This instrument indicates the pitch and bank of the aircraft, which is the tilt of the nose and the wings. The artificial horizon becomes relatively easy to read and interpret once you visualize it from the pilot seat.

The artificial horizon utilizes a small airplane or wings and a horizon bar. The aircraft in the instrument is representative of the host aircraft that is being flown. The horizon bar represents the actual horizon of the Earth. Many artificial-horizon instruments are blue in the upper half representing the sky and dark on the lower half, which represents the ground. If the aircraft or wings in the instrument are above the horizon bar, the aircraft is nose up or climbing. If the aircraft is shown below the horizon bar in the instrument, then the aircraft is nose down, or descending in altitude. When the symbolic aircraft is even with the horizon bar, the plane is in level flight.

The artificial-horizon instrument typically will have degree marks that indicate the amount of aircraft banking that occurs during turning maneuvers. The horizon bar will tilt opposite of the direction of the wings, which will show the orientation of the aircraft to the horizon during banking.

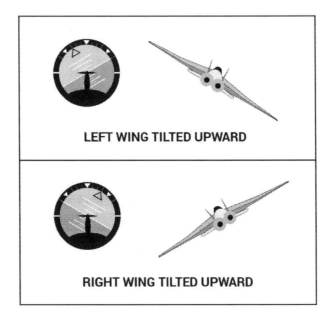

The questions in this section will provide compass and artificial-horizon indications instruments (see example below), and you will be required to determine the orientation of the aircraft using these two. Besides looking at the compass for direction, pay close attention to the artificial horizon to determine if the indication is that of a climbing or descending aircraft, or possibly an aircraft in level flight. Also it is important to verify the direction of banking. Once you have figured out the heading, the pitch, and the banking, you should be able to choose the correct answer from the simulated aircraft images provided.

For the purpose of these questions, the simulated aircraft will appear to be flying away from your view for north, facing you for south, facing with the nose to left of the page for west and to the right for east.

Example practice question: Looking at the instruments on the left, which choice depicts the orientation of the aircraft?

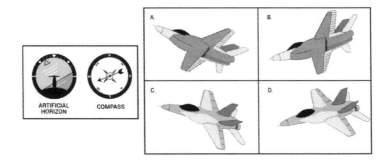

The correct answer to the example question is "A". The compass is showing a west-southwest heading. The artificial horizon shows the wings are above the horizon bar, indicating the aircraft is in a climb. Lastly, the artificial horizon shows the right wing is dipped, as the horizon bar is tilted opposite. This indicates the aircraft is banked to the right. Put it all together and you have an aircraft heading west-southwest (facing away and to the left), while it is in a climb and a right bank. The only choice that corresponds to this orientation is answer A.

Studying for the compass and artificial horizon portion of the test

You may want to practice with a standard compass if you are not familiar with compass reading. Many smartphones have compass capabilities, or you can find one at a sporting-goods store. You probably will not be able to find an artificial horizon to practice with, so it is recommended that you prepare for this portion of the test using the following practice questions.

Practice Questions

1. Looking at the instruments on the left, which choice depicts the orientation of the aircraft?

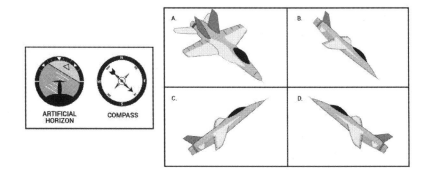

2. Looking at the instruments on the left, which choice depicts the orientation of the aircraft?

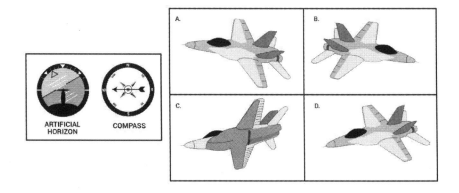

3. Looking at the instruments on the left, which choice depicts the orientation of the aircraft?

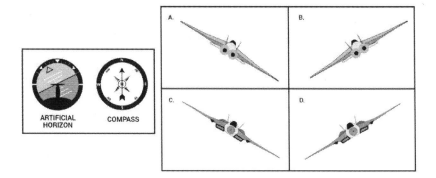

4. Looking at the instruments on the left, which choice depicts the orientation of the aircraft?

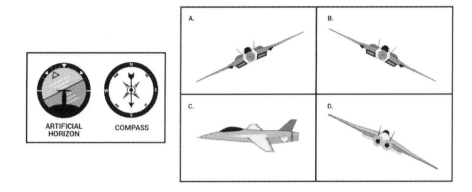

5. Looking at the instruments on the left, which choice depicts the orientation of the aircraft?

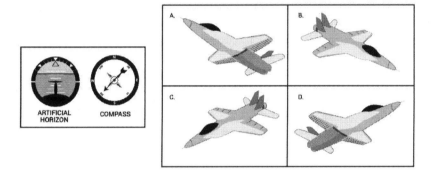

6. Looking at the instruments on the left, which choice depicts the orientation of the aircraft?

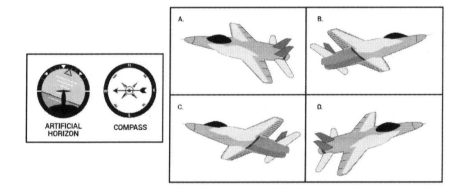

7. Looking at the instruments on the left, which choice depicts the orientation of the aircraft?

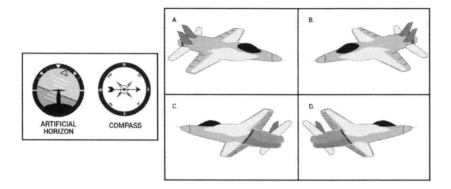

8. Looking at the instruments on the left, which choice depicts the orientation of the aircraft?

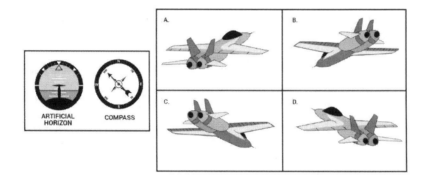

9. Looking at the instruments on the left, which choice depicts the orientation of the aircraft?

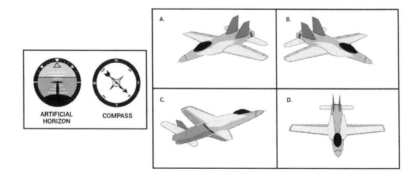

10. Looking at the instruments on the left, which choice depicts the orientation of the aircraft?

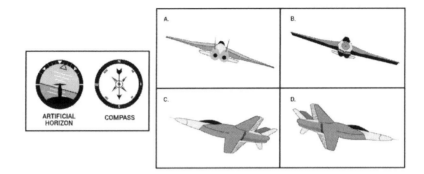

Answers Explanations

1. B: The aircraft is heading southeast, descending, and banking left.

2. C: The aircraft is heading west, descending, and banking right.

3. A: The aircraft is heading north, in level flight, and banking right.

4. A: The aircraft is heading south, in level flight, and banking to the right.

5. C: The aircraft is heading southwest, descending, and without banking.

6. A: The aircraft is heading west, ascending, and banking left.

7. D: The aircraft is heading east, ascending, and banking left.

8. D: The aircraft is heading northwest, ascending, and without banking.

9. B: The aircraft is heading southeast, descending, and without banking.

10. B: The aircraft is heading south, ascending, and banking left.

Aviation Information

The aviation information contained in this section can help provide some basic knowledge in the following areas:

- Fixed-wing aircraft
- Flight envelope
- Flight concepts and terminology
- Flight maneuvers
- Helicopters
- Airport information

Fixed-wing aircraft

A fixed-wing aircraft is one in which movement of the wings in relation to the aircraft is not used to generate lift, even though technically they flex in flight, as do all wings. In contrast, helicopters generate lift through rotating airfoils.

Fixed-wing aircraft have the following items in common: wings, flight controls, tail assembly, landing gear, fuselage, and an engine or powerplant. Here's an illustration of many of the parts of planes:

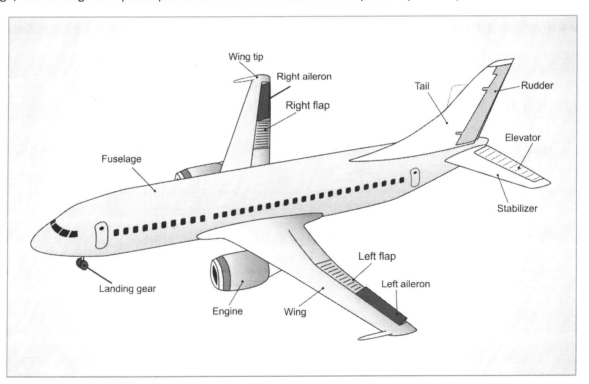

Wings

An airfoil is a surface such as a wing, elevator, or aileron that is designed to help create lift and control of an aircraft by using and manipulating air flow. The standard wing, or airfoil, is mounted to the sides of the fuselage. The outline or cross section of the airfoil is referred to as the airfoil profile. The standard wing has an upper surface that is curved while the lower surface is relatively flat. The air moves over the curved surface of the wing at a higher speed than it moves under the flat surface. The difference in airflow on the surfaces of the wing creates lift due to the aircraft's forward airspeed, which enables flight.

Many modern aircraft structures use full cantilever wings, meaning no external bracing is required. The strength in this design is from the internal structural members and the fuselage. Semi-cantilever wings use some form of external bracing, whether it be wires or struts.

Aircraft wing designs vary based on the characteristics and type of performance needed for the aircraft. Changing the wing design changes the amount of lift that can be created and the amount of stability and control the aircraft will have at different speeds. Some aircraft have wings that are designed to tilt, sweep, or fold. So long as they do not generate lift, they are still considered fixed-wing aircraft. Wing geometry is related to the shape of the aircraft wings when viewed from the front of the aircraft, or from above.

Flight Controls
Flight controls, like the name implies, are used to direct the forces on an aircraft in flight for the purpose of directional and attitude control. Flight controls vary from simple mechanical (manually operated) systems, to hydro-mechanical systems, to fly-by-wire systems. Fly-by-wire systems send signal via wire to control the plane.

When discussing flight controls, it's important to understand the terms *leading edge* and *trailing edge*. *Leading edge* refers to the front part of the wing that separates air, forcing it to go above or below the wing. *Trailing edge* is the back part of the wing and where the air comes back together.

The flight controls typically are divided into primary and secondary systems.

The primary flight controls are responsible for the movement of the aircraft along its three axes of flight. Primary flight controls on an aircraft are the *elevators*, *ailerons*, and *rudder*. Elevators are mounted on the trailing edges of horizontal stabilizers and are used for controlling aircraft pitch about the lateral axis. The ailerons are mounted on the trailing edges of the wings and are used for controlling aircraft roll about the longitudinal axis. The rudder is mounted on the trailing edge of the vertical fin and is used for controlling rotation (yaw) around the vertical axis. Here's an illustration of pitch, roll, and yaw along the three axes.

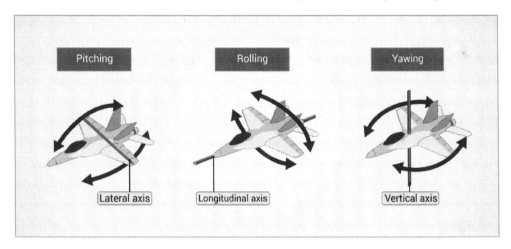

Secondary, or auxiliary, flight controls include (but are not limited to): flaps, slats, spoilers, speed brakes, and the trim system. Flaps are the hinged portion of the wing trailing edge between the ailerons and the fuselage, and sometimes may be located on the leading edge of a wing as well. If used during takeoff, flaps reduce the amount of runway and time needed to takeoff. During landings, flaps increase the drag on the wings slowing the plane down and allowing it to go slower right before it lands, which then reduces the amount of runway needed. Essentially, flaps allow the aircraft to produce more lift at slower airspeeds. Flaps are also utilized on some planes to increase maneuverability. The amount of flap extension and the angle can be adjusted by the flap levers located in the cockpit.

Spoilers are located on the upper, or trailing edge, of the wing and are used to decrease lift. They allow the nose of the aircraft to be pitched down without increasing the airspeed, allowing for a safe landing speed. Slats are located on the middle to outboard portion of the leading edge of the wing. They are used to create additional lift and slower flight by extending the shape of the wing. Flaps and slats are hidden inside the wing and are extended during takeoff and landing.

The trim system is an inexpensive autopilot that eliminates the need for the pilot to constantly maintain pressure on the controls. Most aircraft trim systems contain trim, balance, anti-balance, and servo tabs. These tabs are found on the trailing edge of primary flight control surfaces and are used to counteract hydro-mechanical and aerodynamic forces acting on the aircraft.

Tail Assembly
The tail assembly of an aircraft is commonly referred to as the empennage. The empennage structure typically includes a tail cone, fixed stabilizers (horizontal and vertical), and moveable surfaces to assist in directional control. The moveable surfaces are the rudder, which is attached to the vertical stabilizer, and the elevators attached to the horizontal stabilizers.

Landing Gear
Aircraft require landing gear not only for takeoffs and landings but also to support the aircraft while it is on the ground. The landing gear must be designed to support the entire weight of the aircraft and handle the loads placed on it during landing, as well as be as light as possible. Small aircraft that fly at low-speeds usually have fixed landing gear, which means it is stationary and does not have the ability to retract in flight. Aircraft that fly at higher speeds require retractable landing gear so that the gear is not in the airstream. Retractable landing gear makes the aircraft more aerodynamic by reducing drag, but usually at the cost of additional weight when compared to the fixed landing gear. Also, the landing gear should only be operated (extended or retracted) when the airspeed indicator is at or below the aircraft's maximum landing-gear operating speed, or V_{LO}. Airspeeds above V_{LO} can damage the landing gear operating mechanism. When the gear is down and locked, the aircraft should not be operated above the aircraft's maximum landing-gear extended speed, or V_{LE}. Since landing gear is more stable when all the way down than when being moved, V_{LE} will be higher for an aircraft than V_{LO}. A switch or lever that resembles the shape of a wheel is used to raise and lower the landing gear.

Landing gear is made from a variety of materials, such as magnesium, steel, and aluminum. Most landing gear has some type of shock absorbers and braking system. Not all landing gear uses wheels. Depending on what the aircraft is used for, it may have skis, skids, pontoons, or floats instead of tires for landing on snow, ice, or water.

Landing gear is typically found in two configurations: tricycle gear and conventional gear (also known as tail wheel gear).

Tricycle gear is the most common configuration. Tricycle gear has a single wheel in the front (usually under the nose) and two wheels side-by-side at the center of gravity of the aircraft. Large, heavy aircraft may contain extra wheels in this configuration. Small aircraft equipped with nose landing gear can typically be steered with rudder pedals.

Conventional tail wheel aircraft or "tail draggers" were more common in the early days of aviation. The two main wheels supported the heaviest portion of the aircraft, with the third smaller wheel was near the tail. Having the smaller wheel in the tail allowed the aircraft to rest at an incline on the ground, which provided more clearance for the nose propeller. In some conventional gear, the tail wheel could be steered mechanically with the rudder pedals.

Fuselage
The fuselage is the main structure, or airframe, of an aircraft. The fuselage is where the cockpit, or cabin, of the aircraft is, and where passengers or cargo may be located. It provides attachment points for the main

components of the aircraft such as the wings, engines, and empennage. If an aircraft is a single-engine, the fuselage houses the powerplant as well. If the powerplant is a reciprocating engine, it is mounted in the front of the aircraft, while a turbine engine would be mounted in the rear of the aircraft. *Cowling* is the term for the covering of an airplane's engine which is streamlined to maximize aerodynamics.

Aircraft contain a battery that is charged by an alternator or generator. Wiring takes the electric current to where it's needed throughout the plane. Some type of warning system will be in place to warn of inadequate current output. This could be a warning light or an ammeter. If the number showing on an ammeter is positive, it means the battery is charging. If the number is negative, it means that the alternator or generator cannot keep up with the amount of current being used. Some aircraft have a loadmeter, which shows the amount of current being drawn from the battery.

Fuselage structures are usually classified as truss, monocoque, and semi-monocoque. The truss fuselage is typically made of steel tubing welded together, which enables the structure to handle tension and compression loads. In lighter aircraft, an aluminum alloy may be used along with cross-bracing. A single shell fuselage is referred to as monocoque, which uses a stronger skin to handle the tension and compression loads. There is also a semi-monocoque fuselage, which is basically a combination of the truss and monocoque fuselage, and is the most commonly used.

Powerplant
The powerplant of an aircraft is its engine, which is a component of the propulsion system that generates mechanical power and thrust. Most modern aircraft engines are typically either turbine or piston engines.

Flight Envelope

The flight envelope (also known as the performance envelope or service envelope) refers to the capabilities of an aircraft based on its design in terms of altitude, airspeed, loading factors, and maneuverability. When an aircraft is pushed to the point it exceeds design limitations for that specific aircraft, it is considered to be operating outside the envelope, which is considered dangerous.

All aircraft have approved flight manuals that contain the flight limitations or parameters. To ensure an aircraft is being operated properly, a pilot needs to be familiar with the aircraft's flight envelope prior to flight. The flight parameters are based on the engine and wing design and include the following: maximum and minimum speed, stall speed, climb rate, glide ratio, maximum altitude, and the maximum amount of gravity forces (g-forces) the aircraft can withstand.

- The *maximum speed* of an aircraft is based on air resistance getting lower at higher altitudes, to a point where increased altitude no longer increases maximum speed due to lack of oxygen to "feed" the engine.

- The stalling speed is the minimum speed at which an aircraft can maintain level flight. As the aircraft gains altitude, the stall speed increases (since the aircraft's weight can be better supported through speed).

- The *climb rate* is the vertical speed of an aircraft, which is the increase in altitude in respect to time.

- The *climb gradient* is the ratio of the increase in altitude to the horizontal air distance.

- The *glide ratio* is the ratio of horizontal distance traveled per rate of fall.

- The *maximum altitude* of an aircraft is also referred to as the service ceiling. The ceiling is usually determined by the aircraft performance and the wings, and is where an altitude at a given speed can no longer be increased at level flight.

- The *maximum g-forces* each aircraft can withstand varies, but is based on its design and structural strength.

Commercial aircraft are considered to have a small flight envelope, since the range of speed and maneuverability is rather limited, and they are designed to operate efficiently under moderate conditions. The Federal Aviation Administration (FAA) is the controlling body in the United States pertaining to authorized flight envelopes and restrictions for commercial and civilian aircraft. The FAA may reduce a flight envelope for added safety as needed.

Military aircraft, especially fighter jets, have extremely large flight envelopes. By design, these aircraft are very maneuverable and can operate at high speeds as their purpose requires. The term "pushing the envelope" originally referred to military pilots taking an aircraft to the extreme limits of their capabilities, mostly during combat, but also during aircraft flight testing. The term "outside the envelope" is when the aircraft is pushed outside the design specifications and is considered very dangerous. Operating outside the limits of the aircraft can severely degrade the life of components, or even lead to mechanical failure.

Some of the modern fly-by-wire aircraft have built in flight envelope protection. This protection is built into the control system and helps prevent a pilot from forcing an aircraft into a situation that exceeds its structural and aerodynamic operational limits. This system is beneficial in emergency situations because it prevents the pilots from endangering the aircraft while making split-second decisions.

Flight Concepts and Terminology

Aerodynamics is the study of objects in motion through the air, such as when the air interacts with an aircraft wing, and the forces that produce or change such motion. It is important to note that the air affects the aerodynamics of an aircraft in flight. Therefore, it is important that the aircraft's operating environment, the atmosphere, is understood.

Atmospheric pressure differs depending on altitude. The higher the altitude above sea level, the less pressure exists. Air density varies with pressure. The higher the altitude, the less dense the air. Humidity, the amount of water vapor in the air, varies with the temperature.

There are four main forces that act on all aircraft: *gravity*, *lift*, *thrust*, and *drag*. The four main forces are always present during aircraft flight. For flight to occur, thrust and lift always need to be greater than the gravity and drag forces.

Gravity
Gravity is the downward force that pulls everything toward the surface of the Earth at a rate of 9.8 m/s^2. The weight of an object's mass is the result of gravity acting upon the object. The pull is also called the weight force. Aircraft need to provide enough lift force to overcome gravity to fly.

Lift
The *lift* force is created by the motion of the aircraft through the air. The air that travels over the curved surface of a wing has farther to go than the air moving below the wing. The air above must travel faster than the air below to reach the back of the wing at the same time. The air traveling faster over the top of a wing creates lower pressure. The lower pressure above the wing causes it to lift. *Bernoulli's principle* illustrates that an increase in velocity is always accompanied by a decrease in pressure and creates lift on an airfoil.

<u>Thrust</u>

The engine, or powerplant, of the aircraft creates thrust, which is necessary to create forward motion. *Thrust* is the force that propels the aircraft forward to provide airflow beneath the wings for lift. The purpose of thrust is not to lift an aircraft, but to overcome the force of drag. The direction of thrust can be varied by the design of the aircraft propulsion system.

<u>Drag</u>

Drag is the force generated when an aircraft is moving through the air. Drag is air resistance that opposes thrust, basically aerodynamic friction or wind resistance. The amount of drag is dependent on several factors, including the shape of the aircraft, the speed it is traveling, and the density of the air it is passing through. There are two categories of drag: *induced drag* and *parasitic drag*. Induced drag is created through the generation of lift and is caused by wingtip vortices (rotating air coming off wings). Parasitic drag is created by the shape of the aircraft.

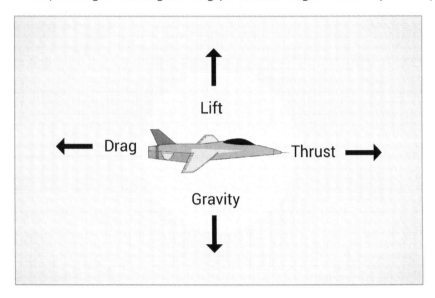

The angular difference measured between an aircraft's axis and the line of the horizon is the flight attitude. Aircraft attitude is based on relative positions of the nose and wings on the natural horizon. There are four types of attitude flying: pitch control, bank control, power control, and trim control. The angle formed by the longitudinal axis of the aircraft is pitch attitude. The angle formed by the lateral axis is bank attitude. Yaw is the rotation of the aircraft around its vertical axis, which is not relative to the horizon, but to the flightpath. Power control is used when there is a need for a change in thrust in the aircraft. Trim is for relieving all possible control pressures held after the desired attitude has been reached. The moveable surfaces on the wings and tail introduced earlier allow for control of an aircraft's attitude and orientation. These airfoils work on the same principle as the lift on the wing.

Here are some examples of pitch control and bank control:

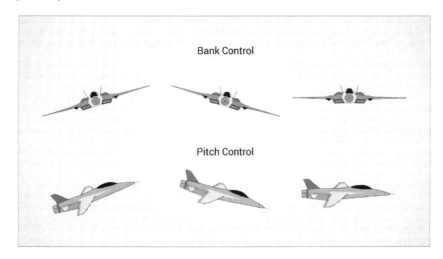

Flight Controls and Forces

The vertical stabilizer, or the tail fin, keeps the aircraft aligned in the desired direction. Air presses on both sides of the tail with equal force when operating in a straight line. When the aircraft yaws left or right, the air pressure increases on one side of the tail and decreases on the other side. The imbalance of the pressure acting upon the tail will push it back in line.

The horizontal stabilizer provides for leveling of an aircraft in flight. If the aircraft tilts up or down, the air pressure increases on one side of the stabilizer and decreases on the other. This imbalance on the stabilizer will push the aircraft back into level flight. The horizontal stabilizer holds the tail down as well, since most aircraft designs induce a tendency for the nose to tilt down due to the center of gravity being forward of the center of lift.

Ailerons are located on the outer, trailing edges of the wings, and control the roll of an aircraft. The ailerons operate opposite each other, up and down, to decrease lift on one wing and increase lift on the other. The change in lift will cause the aircraft to roll either to the left or right; therefore, the pilot will use the ailerons to tilt the aircraft in the direction of a turn.

The elevators are moveable control surfaces attached to the trailing edge of the horizontal stabilizer. The elevators tilt up and down to increase or decrease lift on the tail. The increase or decrease in lift will cause the nose of the aircraft to tilt up or down.

The rudder is a moveable control surface located on the aft end of the aircraft vertical tail fin. The rudder moves from side to side, pushing the tail left or right. The rudder is used in conjunction with the ailerons to turn an aircraft.

Trim Control

In-flight variable trim control along the three axes is used to level an aircraft in pitch and roll while eliminating yaw. The trim system serves to counter the aerodynamic forces that affect the balance of the aircraft in flight. The control pressures on an aircraft change as conditions change during a flight. Trim control systems can compensate for the change in weight and center of gravity that occurs in flight as the aircraft burns fuel, as well as compensating for change in wind and atmospheric pressure. Trim can relieve all possible control pressures held after the desired attitude has been reached. Larger and more sophisticated aircraft tend to have a more sophisticated trim system in the pitch, roll, and yaw axes. Smaller aircraft tend to have only pitch control.

The primary flight controls are positioned in-line with the aircraft surfaces during straight-and-level flight. When the flight is not balanced, one or more flight control surfaces will need to be adjusted by continuous control input. This input can be performed through the use of trim tabs; therefore, avoiding the need for constant adjustments by the pilot. The trim systems also prevent the pilot from having to exert strong control force for an extended amount of time—essentially easing the physical workload required to control the aircraft.

Angle of Attack

The angle of attack (AOA) is the angle between the relative wind and the chord line (line between the leading and trailing edge of the airfoil). The AOA can be used to affect the amount of lift on an aircraft. The greater the AOA, the greater the lift generated until the AOA reaches the critical angle of attack. The critical angle of attack induces a stall because the induced drag exceeds the lift. During a stall, the wing is no longer able to create sufficient lift to oppose gravity. Stall angle is usually around 20°. Thus once it's been reached, any subsequent increase in the AOA hurts aerodynamics.

Additional Related Terms

Center of gravity: In flight, the aircraft rotates around its center of gravity.

Ground effect: The reduced drag and increased lift that is experienced by an aircraft flying in close proximity to the ground is called the ground effect. For fixed-wing aircraft it is considered in ground effect when the wing is in close proximity of the ground. A helicopter is considered to be in ground effect when it is within one rotor diameter of the ground. For example, if the rotor diameter of the helicopter is 50 feet, then 50 feet and below is considered to be in ground effect.

Flight Maneuvers

There are four fundamental maneuvers in flight: straight-and-level, turns, climbs, and descents. Every controlled flight usually includes a combination of these four fundamentals.

The duration and the amount of pressure used on the flight controls determine the displacement of the corresponding flight control, which controls the maneuver being made. Since the airspeed differs during maneuvers, the distance the flight controls move is not as important as the appropriate pressure the pilot applies to the control for each maneuver.

Straight-and-Level Flight

Maintaining a heading and altitude, with only minor adjustments to flight controls, is referred to as straight-and-level flight. Straight-and-level flight may require adjustment depending on the atmospheric conditions — just as keeping an automobile going in a straight line can require adjustments. This laterally level flight can be achieved by checking the relationship of the aircraft wingtips with the natural horizon. The distance of both wingtips above or below the horizon should be the same. Any adjustments should be made with the ailerons. To a lesser extent, the wingtips can also be used to determine pitch attitude. During straight-and-level flight, small deviations from lateral flight can be addressed quickly with small corrections. Almost no control pressure is required to maintain straight-and-level if the aircraft is properly trimmed and is in smooth air. Airspeed will usually remain consistent when a constant power setting is used.

Turns

A turn in an aircraft is performed by banking the wings in the direction of the turn. The appropriate amount of control pressures need to be applied by the pilot to achieve and hold the desired banking for the turn. All primary controls are required and should be coordinated during a turn.

Turns are divided into three classes: shallow, medium, and steep. Shallow turns are typically less than 20°, which is so shallow that the aircraft's lateral stability acts to level the wings, unless an aileron is used to maintain banking. Medium turns range from approximately 20° to 45°, and the aircraft holds a constant bank. Steep turns are more than 45°, and the aircraft will have a tendency to overbank without the addition of aileron control.

During constant altitude and airspeed turns, the elevator should be up to increase the angle-of-attack (AOA) of the wing when rolling into a turn. Use of the elevator during a turn is needed because the vertical lift has been changed to a horizontal lift. To stop the turn, the aileron and rudder are used in the opposite direction to return the wings to level flight.

Climbs

An aircraft's climb is restricted by the amount of thrust available, since the thrust needs to be able to overcome the increased drag that occurs during climbing maneuvers. In a climb, the weight is no longer perpendicular to the flightpath; it is directed rearward.

Climbs require the simultaneous increase in throttle and back-pressure on the elevator to lift the nose of the aircraft. Nose-up elevator trim may be used after the climb has been established to make minor adjustments in pitch.

Descents

Descent is when an aircraft changes flight path to a downward inclined plane, normally losing altitude with partial power. The settings recommended by the aircraft manufacturer should be used when setting pitch and power for a descent. The power, pitch, and airspeed should be kept constant.

Helicopters

Helicopters, a type of rotary-wing or rotorcraft are extremely versatile aircraft that may be used for a wide array of situations where a typical aircraft could not perform. A helicopter differs from other types of aircraft that derive lift from their wings, as it gets its lift and thrust from the rotors. The rotors are basically rotating airfoils; hence the name rotorcraft. The "rotating wing," or main rotor, may have two or more blades whose profiles resemble that of aircraft wings.

Although rotorcraft come in many shapes and sizes, they mostly have the same major components in common. There is a cabin, an airframe, landing gear, a powerplant, a transmission, and an anti-torque system. The cabin of the rotorcraft houses the crew and cargo. The airframe is the fuselage where components are mounted and attached. The landing gear may be made up of wheels, skis, skids, or floats.

Due to its operating characteristics, a helicopter is able to:

- Take off and land vertically (in almost any small, clear area)
- Fly forward, backward, and laterally
- Operate at lower speeds
- Hover for extended periods of time

Common uses for the helicopter include military operations, search and rescue, law enforcement, firefighting, medical transport, news reporting, and tourism.

Helicopters are subjected to the same forces as other aircraft: lift, weight, thrust, and drag. In addition, helicopters also have some unique forces they are subjected to: torque, centrifugal and centripetal force, gyroscopic precession, dissymmetry of lift, effective translational lift, transverse flow effect, and Coriolis effect.

Here are some examples of different helicopter designs:

Torque
Torque from the engine turning the main rotor forces the body of the helicopter in the opposite direction. Most helicopters use a tail rotor, which counters this torque force by pushing or pulling against the tail.

Centrifugal and Centripetal Force
Centrifugal force is the force that causes rotating bodies to move away from the center of rotation. Centripetal force is the force that counteracts centrifugal force, since it keeps an object a certain distance from the center of rotation.

Gyroscopic Precession
Gyroscopic precession is when the applied force to a rotating object is shown 90º later than where the force was applied.

Dissymmetry of Lift
Dissymmetry of lift is the difference in lift between the advancing and retreating blades of the rotor system. The difference in lift is due to directional flight. The rotor system compensates for dissymmetry of lift with blade flapping, which allows the blades to twist and lift in order to balance the advancing and retreating blades. The pilot also compensates by cyclic feathering. Together, blade flapping and cyclic feathering eliminate dissymmetry of lift.

Effective Translational Lift
Effective translational lift (ETL) is the improved efficiency that results from directional flight. Efficiency is gained when the helicopter moves forward, as opposed to hovering. The incoming airflow essentially pushes the turbulent air behind the helicopter, which provides a more horizontal airflow for the airfoil to move through. This typically occurs between 16 and 24 knots.

Transverse Flow Effect
Transverse flow effect is the difference in airflow between the forward and aft portions of the rotor disk. During forward flight, there is more downwash in the rear portion of the rotor disk. The downward flow on the rear portion of the disk results in a reduced AOA and produces less lift. The front half of the rotor produces more lift due to an increased AOA. Transverse flow occurs between 10 and 20 knots and can be easily recognized by the increased vibrations during take-off and landing.

Coriolis Effect
The Coriolis Effect is when an object moving in a rotating system experiences an inertial force (Coriolis) acting perpendicular to the direction of motion and the axis of rotation. In a clockwise rotation, the force acts to the left of the motion of an object. If the rotation is counterclockwise, the force acts to the right of the motion of an object.

A typical helicopter is controlled through use of the cyclic control stick, the collective controls lever, and the anti-torque pedals.

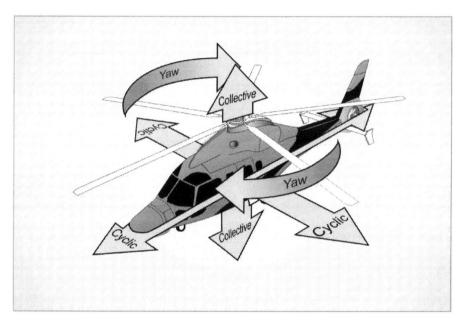

Cyclic Control

The cyclic is the control stick that is used to control the movement of the helicopter forward, backward, or sideways. Simply stated, the cyclic changes the pitch of the rotor blades cyclically in any of 360°. The cyclic stick also contains several buttons that control the trim, intercom, and radio. These will differ slightly depending on each model helicopter.

Collective Controls

The collective control is used to increase the pitch of the main rotor simultaneously at all points of the rotor blade rotation. The collective increases or decreases total rotor thrust; the cyclic changes the direction of rotor thrust. In forward flight, the collective pitch changes the amount of thrust, which in turn can change the speed or altitude based on the use of the cyclic. During hovering, the collective control is used to adjust the altitude of hover.

Anti-Torque Pedals

The anti-torque pedals are used to control yaw on a helicopter. These pedals are located in the same place, and serve a similar purpose, as rudder pedals on an airplane. Application of pressure on a pedal changes the pitch of the tail rotor blade, which will increase or reduce the tail rotor thrust and cause the nose of the helicopter to yaw in the desired direction.

Airport Information

Airports vary in size and complexity, from small dirt or grass strips to a large major airport with miles of paved runways and taxiways. Pilots are required to know the procedures and rules of the specific airports being used, and understand pavement markings, lights, and signs that provide takeoff, landing, and taxiway information.

Procedures have been developed for airport traffic patterns and traffic control. These procedures include specific routes or specific runways for takeoffs and landings. Each airport traffic pattern depends on a number of factors, including obstructions and wind conditions.

<u>Terms</u>

Clearway: A specified area after the runway that is clear of obstacles, at the same or nearly the same altitude as the runway, and specifically for planes to fly over in their initial ascent.

Decision height (DH): The lowest height (in feet) in which if a pilot cannot see specified visual references, they must stop landing

Holding/Flying a Hold: Refers to a plane flying in an oval flight path near the airport while waiting for clearance to land

Runway visual range (RVR): The distance away from the airport in which a pilot should be able to see runway markings and/or lights.

Threshold: The start or end of a runway

Taxiway: The paved area that planes travel on to travel between the terminal and runway. The path a plane should follow on a taxiway is marked by a yellow line

Taxiway intersections: Where two taxiway routes intersect.

<u>Signs</u>

Here are some common taxiway signs:

Name	Sign Color	Letter Color	Function
no entry	red	white	indicates an area planes should not go into
runway location	black	yellow	displays current runway name
taxiway location	black	yellow	displays current taxiway name
direction/runway exit	yellow	black	displays name of upcoming taxiway that airplane is about to intersect with
runway	red	white	displays name of runway that airplane is about to intersect with

<u>Lighting</u>

These lights are visible to about 3 miles in the daytime and up to 20 miles at night.

- *Approach Lighting Systems (ALS)*: A series of light bars leading up to a runway that assist the pilot in lining up with the runway.

- *Runway centerline lights*: There are used to facilitate landing at night or under adverse visibility conditions. These lights are embedded into the runway. They begin as white, then alternate between red and white, and then are completely red at the end of the runway.

- *Obstructions/aircraft warning lights*: *Red* or *white* lights used to mark obstructions (like building and cell towers) both in airports and outside of them.

- *Runway edge lights*: These are *white* and highlight the boundaries of the runway. These are referred to a high, medium, or low intensity depending on the maximum intensity they can produce.

- *Runway end identifier lights (REIL)*: Many airport runways have these two flashing *red* lights. They provide a warning that the runway is ending.

- *Taxiway centerline lights*: *Green* lights indicating the middle of a taxiway.

- *Taxiway edge lights*: *Blue* lights denoting the edge of a taxiway.

- *Threshold lights*: *Green* lights that indicate the start of the runway

Visual Approach Slope Indicators (VASI)

The visual approach slope indicator (VASI) is a light system of two sets of lights designed to provide visual guidance during approach of a runway. The light indicators are visible 3-5 miles away during daylight hours and up to 20 miles away in darkness. The indicators are designed to be used once the aircraft is already visually aligned with the runway. Both sets of lights can appear as white or red. When the front lights appear white and the back lights appear red, it indicates that the plane is at the proper angle. If both lights appear white, it means the pilot is too high, and if both appear red, it means the pilot is too low.

Precision Instrument Runways

Precision instrument runways have operational visual and electronic aids that provide directional guidance. These aides may be the ILS, Precision Approach Radar (PAR), or Microwave Landing System (MLS). Instrument runways have visual aids with a decision height greater than 200 feet and a runway visual range of 2,600 feet. These runways have *runway end lights*, which consists of eight lights (four on each side of the runway) that can appear as green or red—green to approaching planes and red to planes on the runway.

Non-Precision Instrument Runways

Non-precision instrument runways provide horizontal guidance only when there is an approved procedure for a straight-in non-precision instrument approach. The non-precision runways do not have full Instrument Landing System (ILS) capabilities, but they have approved procedures for localizer, Global Positioning System (GPS), Automatic Direction Finder (ADF), and Very High Frequency Omni-Directional Range (VOR) instrument approaches.

Visual Flight Rules Runways (VFR)

Visual flight rules (VFR) runways, also known as visual runways, operate completely under visual approach procedures. The pilot must be able to see the runway to land safely. There are no instrument approach procedures for this type of runway. They are typically found at small airports. Visual runways have centerline, designators, and threshold markings, as well as hold position markings for taxiway intersections.

Localizer Type Directional Aid (LDA)

Directional runways use a Localizer Type Directional Aid (LDA). The LDA provides a localizer-based instrument approach to an airport where, usually due to terrain, the localizer antenna is not in alignment with the runway. The LDA is more of a directional tool to enable the pilot to get close enough to where the runway can be seen.

Practice Questions

1. Which of the following will create lift?
    a. Airflow against the tail section of the aircraft
    b. Faster flow of air over the wing than beneath it
    c. Raised ailerons
    d. Helium in the wings
    e. Thrust from the engine

2. Which of the following axes is controlled by ailerons?
    a. Lateral
    b. Longitudinal
    c. Vertical
    d. Equatorial
    e. Felling axes

3. Which of the following axes is controlled by elevators?
    a. Lateral
    b. Longitudinal
    c. Vertical
    d. Equatorial
    e. All of the above

4. Which of the following axes is controlled by a rudder?
    a. Lateral
    b. Longitudinal
    c. Vertical
    d. Equatorial
    e. None of the above

5. Which of the following is not a secondary or auxiliary flight control?
    a. Flap
    b. Spoiler
    c. Aileron
    d. Slat
    e. All of the above

6. What component is not part of an empennage?
    a. Stabilizer
    b. Rudder
    c. Elevator
    d. Truss
    e. Slats

7. Which of the following is not a type of landing gear?
    a. Ski
    b. Skate
    c. Skid
    d. Float
    e. Pontoon

8. Which is not a type of fuselage?
    a. Monocoque
    b. Semicoque
    c. Semi-Monocoque
    d. Truss
    e. None of the above

9. Who should be the most concerned about an aircraft's flight envelope?
    a. The pilot
    b. Air traffic controller
    c. Passengers
    d. Aircraft mechanic
    e. People on the ground

10. What is the minimum speed at which an aircraft can maintain level flight?
    a. Ceiling speed
    b. Cruising speed
    c. Stalling speed
    d. Mach 1
    e. Gliding speed

11. What is the maximum operating altitude for a design of an aircraft?
    a. Stratosphere roof
    b. Service ceiling
    c. Stratosphere ceiling
    d. Overhead ceiling
    e. Upper limit

12. Which of the following is another word for aerodynamic friction or wind resistance?
    a. Lift
    b. Gravity
    c. Thrust
    d. Drag
    e. Stall

13. What is the purpose of the horizontal stabilizer?
    a. Pushes the tail left or right in line with the aircraft
    b. Levels the aircraft in flight
    c. Controls the roll of an aircraft
    d. Decreases speed for landing
    e. Decreases speed in a turn

14. Where are ailerons located?
    a. The trailing edge of a rudder
    b. The outer leading edge of a wing
    c. The outer trailing edge of a wing
    d. The inner trailing edge of a wing
    e. The empennage

15. What occurs when the induced drag on an aircraft exceeds its lift?
    a. Roll
    b. Yaw
    c. Pitch
    d. Stall
    e. Gravity

16. Which is not considered to be one of the four fundamental flight maneuvers?
    a. Landing
    b. Straight and level
    c. Turns
    d. Descent
    e. Climbing

17. What force acting upon a helicopter attempts to turn the body of the helicopter in the opposite direction of the main rotor travel?
    a. Gyroscopic precession
    b. Centrifugal force
    c. Torque
    d. Coriolis effect
    e. Centripetal force

18. What is used to increase the pitch of the main rotor at the same time at all points of the rotor blade rotation?
    a. Cyclic control
    b. Collective control
    c. Coriolis control
    d. Symmetry control
    e. Anti-torque control

19. Visual approach slope indicators are visible from what distance during clear, daylight hours?
    a. 1-2 miles away
    b. 3-5 miles away
    c. 5-10 miles away
    d. 10-20 miles away
    e. 20-25 miles away

20. Non-precision instrument runways provide what kind of guidance?
    a. Horizontal
    b. Vertical
    c. Locational
    d. Directional
    e. Multi-directional

Answer Explanations

1. B: Faster flow of air over the wing than beneath it. The air moves over the curved surface of the wing at a higher rate of speed than the air moves under the lower flat surface, which creates lift due to the aircraft's forward airspeed and enables flight.

2. B: Longitudinal. The ailerons are mounted on the trailing edges of the wings, and they are used for controlling aircraft roll about the longitudinal axis.

3. A: Lateral. Elevators are mounted on the trailing edges of horizontal stabilizers and are used for controlling aircraft pitch about the lateral axis.

4. C: Vertical. The rudder is mounted on the trailing edge of the vertical fin and is used for controlling rotation (yaw) around the vertical axis.

5. C: Aileron. The aileron is a primary flight control.

6. E: Slats. Slats are part of the wing. The tail assembly of an aircraft is commonly referred to as the empennage. The empennage structure usually includes a tail cone, fixed stabilizers (horizontal and vertical), and moveable surfaces to assist in directional control. The moveable surfaces are the rudder, which is attached to the vertical stabilizer, and the elevators attached to the horizontal stabilizers.

7. B: Skate. Depending on what the aircraft is used for, it may have skis, skids, pontoons, or floats, instead of tires, for landing on ice or water.

8. B: Semicoque. There are two types of fuselage structures: truss and monocoque. The truss fuselage is typically made of steel tubing welded together, which enables the structure to handle tension and compression loads. In lighter aircraft, an aluminum alloy may be used along with cross-bracing. A single-shell fuselage is referred to as monocoque, which uses a stronger skin to handle the tension and compression loads. There is also a semi-monocoque fuselage, which is basically a combination of the truss and monocoque fuselage and is the most commonly used. The semi-monocoque structure includes the use of frame assemblies, bulkheads, formers, longerons, and stringers.

9. A: The pilot. To ensure an aircraft is being operated properly, a pilot needs to be familiar with the aircraft's flight envelope before flying.

10. C: Stalling speed. The stalling speed is the minimum speed at which an aircraft can maintain level flight. As the aircraft gains altitude, the stall speed increases, since the aircraft's weight can be better supported through speed.

11. B: Service ceiling. The maximum altitude of an aircraft is also referred to as the service ceiling. The ceiling is usually decided by the aircraft performance and the wings, and is where an altitude at a given speed can no longer be increased at level flight.

12. D: Drag. Drag is the force generated when aircraft is moving through the air. Drag is air resistance that opposes thrust, basically aerodynamic friction or wind resistance. The amount of drag is dependent upon several factors, including the shape of the aircraft, the speed it is traveling, and the density of the air it is passing through.

13. B: Levels the aircraft in flight. The horizontal stabilizer provides for leveling of aircraft in flight. If the aircraft tilts up or down, air pressure increases on one side of the stabilizer and decreases on the other. This imbalance on the stabilizer will push the aircraft back into level flight. The horizontal stabilizer holds the tail down as well, since most aircraft designs induce the tendency of the nose to tilt downward because the center of gravity is forward of the center of lift in the wings.

14. C: The outer trailing edge of the wing. Ailerons are located on the outer trailing edges of the wings and control the roll of an aircraft.

15. D: Stall. An aircraft stall occurs at the critical AOA, where the induced drag exceeds the lift. During a stall, the wing is no longer able to create sufficient lift to oppose gravity. Stall angle is usually around 20°.

16. A: Landing. There are four fundamental maneuvers in flight: straight-and-level, turns, climbs, and descents. Every controlled flight usually includes a combination of these four fundamentals.

17. C: Torque. Torque from the engine turning the main rotor forces the body of the helicopter in the opposite direction. Most helicopters use a tail rotor, which counters this torque force by pushing or pulling against the tail.

18. B: Collective control. The collective control is used to increase the pitch of the main rotor simultaneously at all points of the rotor blade rotation. The collective increases or decreases total rotor thrust; the cyclic changes the direction of rotor thrust. In forward flight, the collective pitch changes the amount of thrust, which in turn can change the speed or altitude based on the use of the cyclic. During hovering, the collective pitch will alter the hover height.

19. B: 3-5 miles away. Visual approach slope indicator (VASI) is a light system designed to provide visual guidance during approach of a runway. The light indicators are visible 3-5 miles away during daylight hours and up to 20 miles away in darkness. The indicators are designed to be used once the aircraft is already visually aligned with the runway.

20. A: Horizontal. Non-precision instrument runways provide horizontal guidance only when there is an approved procedure for a straight-in non-precision instrument approach.

Hidden Figures

The Hidden Figures section of the AFOQT evaluates a test taker's ability to quickly and accurately assess visual information. Test takers are given eight minutes to complete fifteen questions, each of which entails scanning a somewhat complicated drawing for a smaller embedded hidden figure or shape. Selecting the correct answer requires skill in spatial awareness, template matching, and careful focus under pressure.

The Hidden Figures questions measure where each candidate falls on the Field Independence–Field Dependence Scale. A candidate whose score falls high on the Field Independence end of the scale is able to focus on only the important information in their visual field, while ignoring the irrelevant information. Those whose scores fall close to the Field Dependence end of the scale struggle to hone in on the relevant information and cues in the visual field and get distracted by unimportant information within the field of view. Air Force admissions officers are looking for candidates that have a high degree of Field Independence because it is crucial that Air Force Officers can accurately and nearly instantly interpret visual information. There will certainly be times when Air Force officers must fly in difficult conditions to drop cargo or pick up troop members in precise locations in a dense jungle or snowy mountaintop. In such situations, the ability to quickly scan one's visual field and hone in on important—and perhaps hidden—visual information while ignoring everything that doesn't matter is crucial.

As mentioned, the Hidden Figures section of the AFOQT contains fifteen questions: three sets of five questions. At the beginning of each of the sets, there are five different figures labeled A-E. Each of the fifteen questions depicts a complex arrangement of shapes, patterns, and lines, which actually have one of the five figures (A-E) at the top of the set embedded within it. Test takers must identify and select the one lettered figure that is hidden in the puzzle-like image. It is important to note that the embedded hidden images will always be of the same size and orientation as they were at the top of the section; each question contains exactly one hidden image within it. However, even though there are five labeled images and five questions per set, not all images are necessarily hidden in one of the five questions. Likewise, images may be hidden in more than one question. For example, figure B may be hidden in question 3 and 4, while figure D is not hidden in any of the five questions.

Test takers should note that figures may be well-hidden and require some time to evaluate and locate. Some arrangements and patterns can pose a challenge even for candidates with a high degree of Field Independence. With that said, the good news is that the Hidden Figures section is unscored; answers do not directly impact a test taker's composite score or his or her passing status for general admissions as an Air Force Officer candidate or for the more rigorous navigator and pilot positions. Still, achieving a high score on this section of the AFOQT will demonstrate to Air Force admissions officers that a candidate is strong in these critical skills.

There are a couple of different strategies that can be used on the Hidden Figures questions. Certain strategies may be more appropriate and helpful for certain figures and puzzle arrangements as well as the needs and strengths of each test taker. Test takers should experiment with a variety of techniques to see what approach is most helpful for this section and should bear in mind that some combination of the two may be optimal.

Strategy 1: Some test takers start by examining the five labeled figures at the top of each set, and look for those with distinguishing features such as long straight sides or sharp angles or protrusions. From there, the complex arrangements of each question are evaluated, with a focus on uncovering the labeled figures that had a prominent feature by searching for the prominence. If the arrangement clearly cannot accommodate the prominent feature, that figure can be eliminated as a possibility and the remaining figures can be tried.

Strategy 2: Some test takers prefer to begin by systematically examining each of the puzzle-like arrangements in the set of questions. This strategy also tends to be better when the hidden labeled figures do not have very pronounced features from which they can be identified. While focusing on one arrangement (one question) at a time, the test taker tries to fit each of the labeled figures from the top of the set into the question one by one. Because there is only one figure hidden in each arrangement, as soon as a fit is identified, the answer is marked and they can move on to the next question. Again, it is important to note that figures may appear in more than one arrangement (so test takers should not cross the figure off after it is found in one question) and the figures, when hidden, will be exactly as they appear at the top of the set (same orientation, size, shape, etc.)

Practice Questions

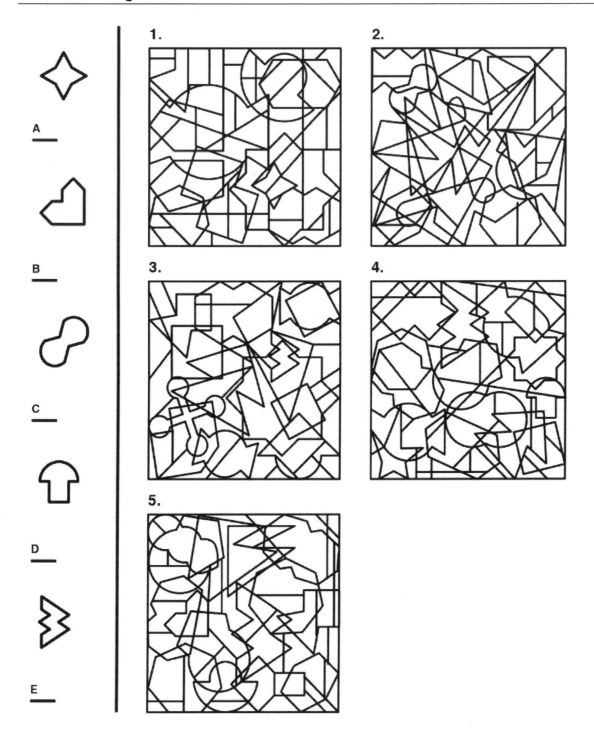

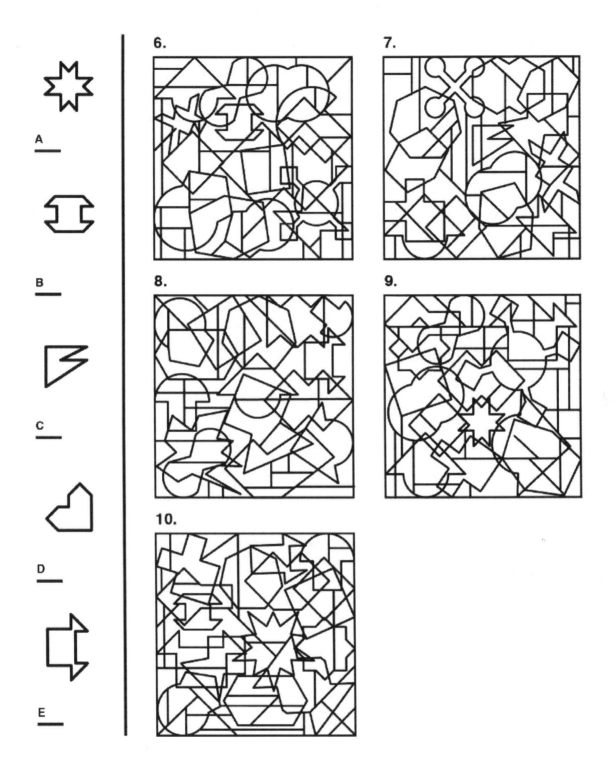

6.

7.

8.

9.

10.

A

B

C

D

E

Answer Explanations

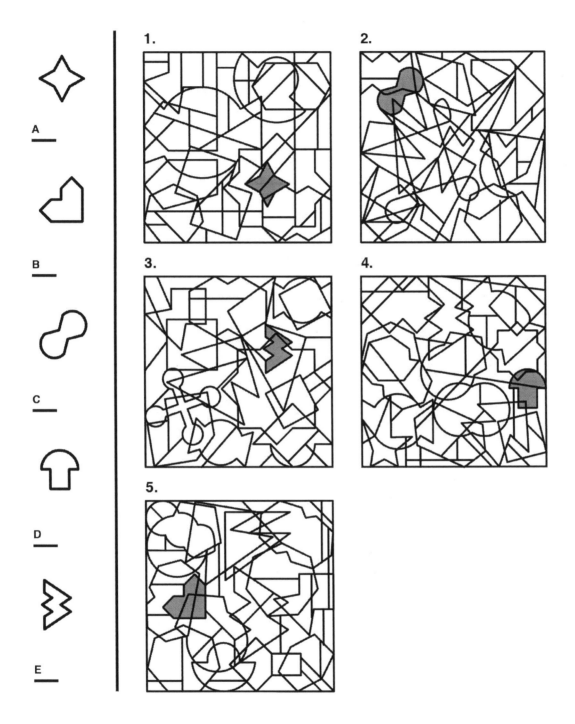

11.

12.

13.

14.

15.

A

B

C

D

E

Rotated Blocks

The Rotated Blocks section of the AFOQT evaluates a test taker's ability to accurately assess visual information in the form of two-dimensional representations of three-dimensional multi-part block arrangements. Test takers are given thirteen minutes to complete fifteen questions, each of which entails examining a depiction of a three-dimensional arrangement of blocks and then identifying which of the five provided answers presents a view from a different angle of the same block arrangement. Selecting the correct answer requires skills in spatial awareness, visual and imaginative ability, structural and design aptitude, and careful focus under pressure.

The challenge presented by the questions in this section is that the differences between the various answer choices are often quite subtle. The options will all look strikingly similar and may only have slight differences in thickness, shape, length, or another structural detail in just a component of the entire arrangement. Test takers will have to imagine the initial multi-part block arrangement in three-dimensional form and then systematically examine the depictions in each of the five answer choices to determine which contains a plausible rotation of the original form.

Some test takers who possess a strong innate ability to visualize three-dimensional figures and manipulate them in their mind's eye into different angles may find this section quite doable. However, most test takers find the task presented in the Rotated Blocks section to be demanding and unnatural. For this reason, dedicated practice is extremely important. With that said, the good news is that the Hidden Figures section is unscored; answers do not directly impact a test taker's composite score or his or her passing status for general admissions as an Air Force Officer candidate or for the more rigorous navigator and pilot positions. Still, achieving a high score on this section of the AFOQT will demonstrate to Air Force admissions officers that a candidate has a high degree of spatial ability—a critical skill for Air Force officers.

There are a couple of different strategies that can be used on the Rotated Blocks questions. Certain strategies may be more appropriate and helpful for certain block arrangements as well as the needs and strengths of each test taker. Test takers should experiment with a variety of techniques to see what approach is most helpful for this section and should bear in mind that some combination of the two may be optimal.

Strategy 1: Some test takers start by examining the provided block arrangement to look for a distinguishing feature such as hole, a long straight side, or a specific protrusion. From there, the rotated arrangement of each question is evaluated, with a focus on uncovering the prominent feature from the initial figure. If the arrangement clearly does not contain the prominent feature or it is positioned in an incorrect location, that choice can be eliminated as a possibility and the remaining arrangements can be addressed. When using this strategy, it is best to pick only one specific detail or prominent feature from the initial arrangement and focus on just this aspect when ruling out the answer choices. Selecting multiple features may cause distraction, confusion, and ultimately, errors.

Strategy 2: Some test takers prefer to begin by systematically examining each of the block arrangements in the answer choices and then eliminate any options that are obviously incorrect. This strategy also tends to be better when the initial multi-part block figures do not have very pronounced features from which they can be identified. While focusing on one arrangement (one answer choice) at a time, the test taker tries to note whether the arrangement is clearly impossible; for example, the blocks may be reversed or missing an obvious section from the original form. After eliminating blatantly incorrect options, the test taker can focus more closely on the remaining choices.

It should be noted that there are approximately 50 seconds to examine and answer each question. This sounds deceptively lenient; test takers must move quickly through each problem. It is prudent for a test taker to skip a question and flag it for review if it poses a significant challenge. Rather than wasting time struggling to find the answer, it is wise to simply mark it and move on, hoping to return to it after the remaining questions have been answered. Sometimes revisiting a question that was previously frustratingly difficult can provide a fresh perspective and a rejuvenated problem-solving ability—helpful ingredients to triumph over challenging arrangements.

Practice Questions

1.

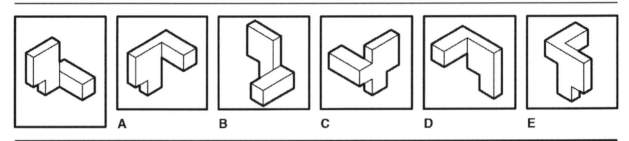

2.

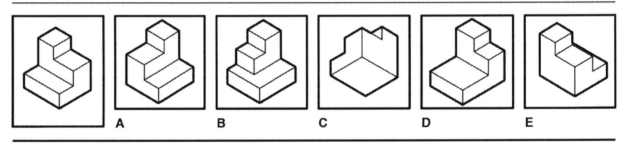

3.

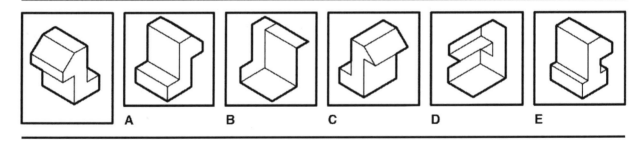

4.

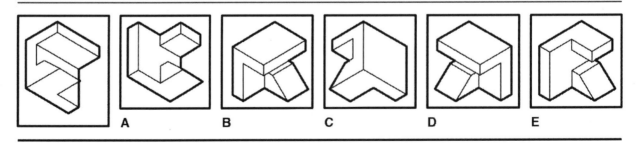

5.

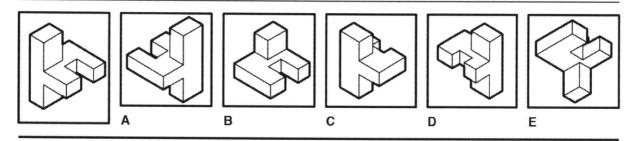

6.

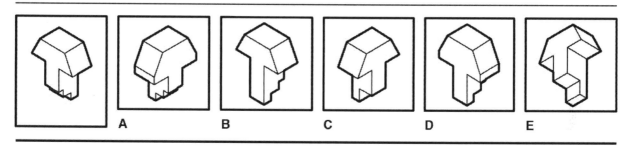

7.

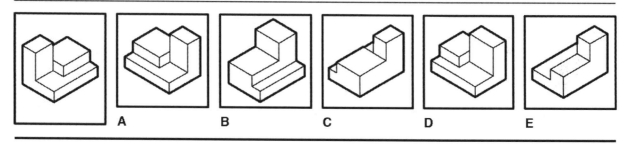

8.

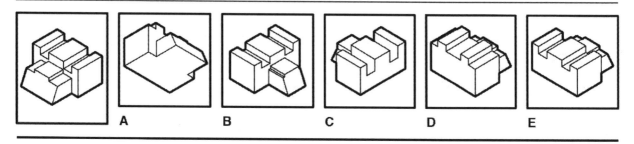

217

9.

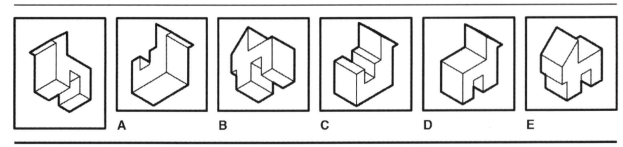

A   B   C   D   E

10.

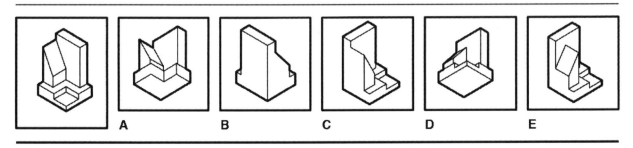

A   B   C   D   E

11.

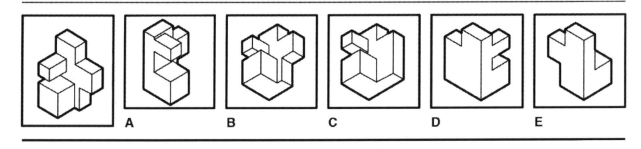

A   B   C   D   E

12.

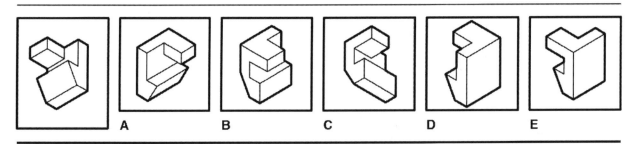

A   B   C   D   E

15.

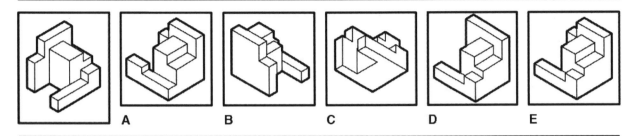

14.

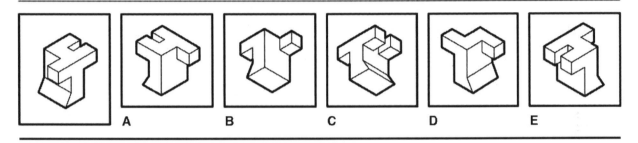

13.

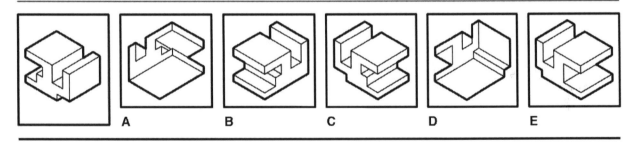

219

Answer Explanations

1. B	6. B	11. B
2. C	7. C	12. D
3. A	8. E	13. C
4. D	9. D	14. E
5. A	10. C	15. E

FREE Test Taking Tips DVD Offer

To help us better serve you, we have developed a Test Taking Tips DVD that we would like to give you for FREE. **This DVD covers world-class test taking tips that you can use to be even more successful when you are taking your test.**

All that we ask is that you email us your feedback about your study guide. Please let us know what you thought about it – whether that is good, bad or indifferent.

To get your **FREE Test Taking Tips DVD**, email freedvd@studyguideteam.com with "FREE DVD" in the subject line and the following information in the body of the email:

> a. The title of your study guide.

> b. Your product rating on a scale of 1-5, with 5 being the highest rating.

> c. Your feedback about the study guide. What did you think of it?

> d. Your full name and shipping address to send your free DVD.

If you have any questions or concerns, please don't hesitate to contact us at freedvd@studyguideteam.com.

Thanks again!

Made in the USA
Middletown, DE
09 February 2018